LAST AID

The Medical Dimensions of Nuclear War

Ground zero before the bombing, Hiroshima, 1945. The circles mark 1000-foot intervals. (U.S. Air Force.)

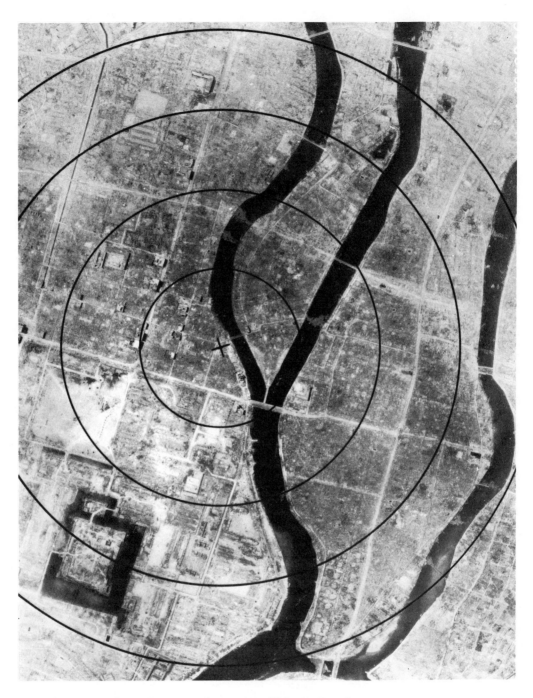

Ground zero after the bombing, Hiroshima, 1945. (U.S. Air Force.)

A nuclear war cannot be won and must never be fought.

President Ronald Reagan
Camp David Address
April 14, 1982

The peoples should know the truth about the consequences, ruinous for mankind, which nuclear war would bring.

Chairman Leonid Brezhnev
26th Communist Party Congress
Moscow
February 23, 1981

LAST AID

The Medical Dimensions of Nuclear War

International Physicians for the Prevention of Nuclear War

Edited by

Eric Chivian, M.D.
Massachusetts Institute of Technology

Susanna Chivian
Institute for Defense and Disarmament Studies

Robert Jay Lifton, M.D.
Yale University

John E. Mack, M.D.
Harvard Medical School

W. H. Freeman and Company
San Francisco

Project Editor: Patricia Brewer; Designer: Gary A. Head;
Production Coordinator: Sarah Segal; Illustration Coordinator:
Richard Quiñones; Artist: J&R Art Services;
Compositor: Allservice Phototypesetting Company;
Printer and Binder: The Maple–Vail Book Manufacturing Group

Library of Congress Cataloging in Publication Data
Main entry under title:

Last aid.

Consists chiefly of papers from the First Congress
of International Physicians for the Prevention of
Nuclear War, held in Washington, D.C., March 1981.
  Bibliography: p.
  Includes index.
  1. Radiation—Toxicology—Congresses. 2. Disaster
medicine—Congresses. 3. Atomic warfare—Congresses.
I. Chivian, Eric. II. International Physicians for the
Prevention of Nuclear War. Congress. (1st : 1981 :
Washington, D.C.)
RA1231.R2L28 1982   616.9'897   82-13472

ISBN 0-7167-1434-5
ISBN 0-7167-1435-3 (pbk.)

Copyright © 1982 by International Physicians for the Prevention of
Nuclear War, Inc.

Printed in the United States of America

1 2 3 4 5 6 7 8 9 0   MP   0 8 9 8 7 6 5 4 3 2

CONTENTS

Section III   Nuclear War, 1980's: The Physical and Medical Consequences   109

Section IV   Nuclear War, 1980's: The Medical Response   179

Medic treating a wounded paratrooper during the first moments of the invasion of Germany. World War II, 1945. (Robert Capa from Magnum for Life, 1945.)

Foreword

In ancient Rome, the doctors who traveled with the armies and looked after illnesses and injuries were called the immunes. These professionals were exempt from public service (*munus*) in the expectation that they would pay strict and exclusive attention to their medical duties. In warfare ever since, doctors have gone about their business as though immune in this antique sense: they are not supposed to carry arms or to get shot at (although my recollection of Okinawa is that they did and were), and they have had very little say, if any, in the technology of war itself. But the technologies involved in repairing traumatic wounds and countering the spread of disease through the world's armies have been developed largely by medical research during times of war. The improvements in life-saving measures were amply demonstrated on the battlefields of both world wars and in Korea and Vietnam, and peacetime society owes much to what was learned. Exempt from other roles in combat, the medical profession was free to devote its energy and intellect to discovering better ways to preserve life and limb in the face of increasingly devastating inventions in weaponry.

From time to time in recent years, the views of nonmilitary physicians have been solicited and listened to by war departments, but only for limited and restricted topics. The decision to give up microbial warfare was perhaps influenced by doctor-advisors, but more because of the wild impracticality of the technology and its likelihood to affect one's own troops than because of any humanitarian considerations. Chemical toxins will, with luck, be put off, maybe for a long time; pharmacology still has too little to offer for protecting either side. But no general staff has as yet taken advice from their medical professionals to give up certain weapons simply because they kill too many people. Up to now, that is what military technologies have been designed to do: the more dead the better.

Now, things have changed. The doctors of the world cannot any longer be professionally immune, exempt, expected to come on the battlefield after the injuries have been inflicted and do their best to fix things up. They know some things for sure about the new weapons.

The essays in this book are intended to reach as many people as possible, here and abroad, in the conviction that people in general are mortally involved. Thermonuclear warfare is not a technical problem for military specialists. It carries lethal menace to human civilization and to the human species itself.

I hope that the essays will be read by some of the political personages and their advisors who are responsible for running the governments of the earth.

These are high hopes for any book but there is another smaller, more exclusive group of readers who should look carefully and thoughtfully at what is written in the chapters to follow. The book is aimed straight at these people. They are the armed services, the soldiery, the professional class responsible for the *profession of arms*. They are, by and large, intelligent people, highly educated in the science and technology of warfare, trained necessarily, by the very nature of their craft and occupation, to think well ahead of the rest of us. The societies they serve cannot survive—on any side—the use of thermonuclear weapons. The politicians can always find expedients in policy, new rhetorical ways to waffle; they can always postpone decisions, hope for the best, maybe serve out their terms. The military caste is in a different situation. Neither they, nor their calling, nor their long tradition of devotion to their countries, can conceivably survive the outright destruction of their societies. What lies ahead for these professionals, if even the neatest and cleanest of nuclear weapons are launched from either side, is not warfare in any familiar sense of the term. It cannot be regarded as organized combat, or techno- logical maneuvering, or anything remotely resembling the old clash of arms. It will be something entirely new, beyond any pro- fessional adroitness in defense, beyond mending at the end. Once begun, there will be no pieces to pick up, no social system to regroup or reorganize, nothing to command.

I hope the generals, the admirals, the captains and the colonels, the senior faculties of all the war colleges, and all the rising young lieutenants in all the armed forces of all the countries will examine the contents of this book with the greatest care. They, and I am beginning to believe they alone, can do the persuading that the world's governments need most quickly: to give up, on all sides, thermonuclear weapons of all kinds, and to make sure that such devices for humankind's suicide are forever abandoned.

Lewis Thomas, M.D.

Preface

This book was in part inspired by the extraordinary development
of the American organization Physicians for Social Responsibility.
It is the direct result of the First Congress of International Physicians for the Prevention of Nuclear War, which was held outside
Washington, D.C., in March 1981 (see Appendix). At this historic
meeting 72 physicians from 12 countries, including leaders in
medicine from the United States, the Soviet Union, the United
Kingdom, Canada, Western Europe, and Japan, put aside their
political and ideological differences to work together to prevent the world's most serious threat to human health and life—
nuclear war.

Experience with nuclear weapons that would allow us to
comprehend modern nuclear warfare is extremely limited. We
learned a great deal from Hiroshima and Nagasaki about the effects of single, small nuclear weapons exploded in the air over
small cities. We also know what happens when the much larger
thermonuclear weapons of our modern arsenals (100 to 1000 times
more powerful than the Hiroshima bomb) are exploded, but these
tests have taken place on coral reefs or at isolated, uninhabited
sites of dirt and sand. It is not possible to predict with certainty
what actually would occur if hundreds or thousands of nuclear
weapons were exploded over major cities in the United States, the

Soviet Union, or other countries. The extent of the firestorms; the levels and distribution of radioactive fallout; the secondary effects from the destruction of the medical care system, of housing, food, and water supplies; and the disruption of the economic, social, political, and environmental fabric on which human life now depends—all these are inadequately understood. What is clear is that nuclear war would result in more death, injury, and disease than any war, natural catastrophe, or epidemic in all history. Life for survivors in a devastated world would be filled with despair, pain, infection, starvation, and mass death.

The focus of this book is medical, because it is perhaps more evident in this area than in any other that plans to protect people from the lethal effects of nuclear weapons, to recover from a nuclear war and to win one, are meaningless. Understanding these basic facts is the first step to an aroused world public opinion and to the outlawing of nuclear weapons from all countries.

Most of the papers in this volume were presented at the First Congress; some resulted from discussions between participants. The remainder were solicited specifically for this book or were taken from various scientific sources to provide a broader view of the subject.

Although we recognize that the effects of even a single modern nuclear weapon on Washington, D.C., or Moscow, or London, or Tokyo would dwarf the tens of thousands of dead, burned, crushed, and irradiated victims of the first atomic bombs, we have examined Hiroshima and Nagasaki because they are our only examples of nuclear destruction on populations. We also have drawn on experience gained from other medical disasters and on the known and anticipated effects of nuclear weapons to estimate the medical consequences of nuclear war and the insoluble problems that would confront surviving physicians and other health care workers.

Most of the photographs used to illustrate this volume are of Hiroshima and Nagasaki, taken from the volume *Hiroshima–Nagasaki: A Pictorial Record of the Atomic Destruction*, published by the Committee of Japanese Citizens to Send Gift Copies of a Photographic and Pictorial Record of the Atomic Bombing to Our Children, and Fellow Human Beings of the World. We are indebted to this group, also called the Hiroshima–Nagasaki Publishing Committee, for permission to use these photographs. Other photographs are taken from various United States government files that record the history of the nuclear weapons testing program. Still others are of conventional warfare. In doing this, we do not mean to imply that conventional weapons effects have any direct relation to nuclear war, but simply that they illustrate, in microcosm, some aspects of nuclear weapons and of the human tragedy of war.

Many people (knowingly and unknowingly) participated in the conception of this volume. In particular, the editors are indebted to George Kistiakowsky, Professor of Chemistry, Emeritus, at Harvard University, former Head of the Manhattan Project's Explosive Division, and Chief Science Advisor to President Eisenhower, for his enormous inspiration over the last two decades to those of us working to prevent nuclear war. And we owe a posthumous debt to Drs. Walter Bradford Cannon and Paul Dudley White, giants in American medicine, who demonstrated the importance of a close working relationship between American and Soviet physicians as a key factor for peace.

Many others worked on the preparation of the materials. John Morris, former Picture Editor of the *New York Times*, and Caren Keshishian, a Washington, D.C., Picture Editor, were responsible for securing several of the photographs. Spectrum Color Labs, Inc., of Boston generously reproduced several photographs. Cybèle Chivian, age 13, provided a fresh, intuitive eye for the selection of several others. Drs. Bernard Lown, Herbert Abrams, and James Muller, co-founders and President, Vice President, and Secretary respectively of International Physicians for the Prevention of Nuclear War, made numerous suggestions about the content of this book with their usual eloquence, insight, and scientific precision. Leslie Kenderes was responsible for typing most of the manuscript. And Linda Chaput, our editor at W. H. Freeman and Company, consistently supported this project from the beginning and saw it through to publication.

To these people, and to the many others who helped us along the way, we are deeply grateful.

June 1982

Eric Chivian, M.D.
Susanna Chivian
Robert Jay Lifton, M.D.
John E. Mack, M.D.

Editors

Eric Chivian, M.D., Staff Psychiatrist at Massachusetts Institute of Technology, is co-founder and Treasurer of International Physicians for the Prevention of Nuclear War. With Drs. Ira Helfand and Helen Caldicott, he revived Physicians for Social Responsibility. He has lectured in the United States and abroad and written widely on the medical and psychiatric aspects of nuclear war.

Susanna Chivian is a freelance editor; she formerly worked for McGraw-Hill Book Company. She is currently an officer of the Institute for Defense and Disarmament Studies.

Robert Jay Lifton, M.D., holds the Foundations' Fund for Research in Psychiatry Professorship at Yale University School of Medicine. He is a member of the National Advisory Board for Physicians for Social Responsibility. His books include *Death in Life: Survivors of Hiroshima* (National Book Award in Science for 1969), *Home from the War,* a study of Vietnam veterans, *History and Human Survival,* and most recently, *The Broken Connection: On Death and the Continuity of Life.* Dr. Lifton is currently completing a psychological study of Nazi doctors.

John E. Mack, M.D., Professor of Psychiatry and Chairman of the Executive Committee of the Departments of Psychiatry at Harvard Medical School, is also on the faculty of the Boston Psychoanalytic Institute and founded the Department of Psychiatry at Cambridge Hospital. He is a member of the American Psychiatric Association task force investigating the psychosocial impact of nuclear developments. His biography, *A Prince of Our Disorder: The Life of T. E. Lawrence,* won the 1977 Pulitzer Prize. His most recent book (written in collaboration with Holly Hickler) is *Vivienne: The Life and Suicide of an Adolescent Girl.*

Contributors

Herbert L. Abrams, M.D., is Philip H. Cook Professor of Radiology at Harvard Medical School and Chairman of Radiology at the Brigham and Women's Hospital and the Sidney Farber Cancer Institute. He is the former Chief of the Department of Radiology at Harvard Medical School. He is a co-founder and Vice President of International Physicians for the Prevention of Nuclear War and is on the Board of Directors of Physicians for Social Responsibility.

Tadatoshi Akiba, Ph.D., is Associate Professor of Mathematics at Tufts University and President of the Foundation for International Understanding. He is also a regular contributor to many Japanese magazines and newspapers, including the *Asahi Journal* and *Computopia*.

Evgueni I. Chazov, M.D., is the USSR Deputy Minister of Health, member of the USSR Academy of Sciences, Director-General of the USSR Cardiological Research Center, member of the Presidium of the USSR Academy of Medical Sciences, and Chairman of the Soviet Committee "Physicians for the Prevention of Nuclear War."

John D. Constable, M.D., is Associate Clinical Professor of Surgery at Harvard Medical School, Visiting Surgeon at the Massachusetts General Hospital, and Chief Consultant in plastic surgery at the Shriners Burns Institute in Boston.

Kai Erikson, Ph.D., is Editor of the Yale Review and Professor of Sociology and American Studies at Yale University. He is the author, among other works, of *Wayward Puritans: A Study in the Sociology of Deviance* (1966) and *Everything in Its Path: Destruction of Community in the Buffalo Creek Flood* (1977).

H. Jack Geiger, M.D., is Arthur C. Logan Professor of Community Medicine and Director of the Program in Health, Medicine and Society in the School of Biomedical Education of the City College of New York. He is on the Board of Directors of Physicians for Social Responsibility. In 1973 he received the Rosenhaus Foundation Award from the American Public Health Association.

Alfred Gellhorn, M.D., is Visiting Professor of Health Policy and Management at the Harvard School of Public Health, where he also chairs the Committee on Community Health. He is Adjunct Professor at both the University of North Carolina School of Medicine and the School of Biomedical Education of the City College of New York. He is former Dean of the University of Pennsylvania Medical School and former Director of the School of Biomedical Education at City College of New York.

Samuel Glasstone, Ph.D., and Philip J. Dolan, Ph.D., compiled and edited the U.S. Department of Defense and U.S. Department of Energy's classic book *The Effects of Nuclear Weapons,* first published in 1957 and now in its third edition.

Angelina K. Guskova, M.D., is Professor and Head of the Department of Biophysics at the Institute of Biophysics of the USSR Ministry of Public Health and a member of the Soviet Committee "Physicians for the Prevention of Nuclear War."

Andrew Haines, M.B., B.S., is a member of the Medical Research Council Epidemiology and Medical Care Unit at Northwich Park Hospital, near London. He is also Senior Lecturer in General Practice in the Department of Community Medicine and General Practice at the Middlesex College Medical School.

Howard H. Hiatt, M.D., is Dean of the Harvard School of Public Health and Professor of Medicine at Harvard Medical School. Before becoming Dean in 1972 he was the Herman L. Blumgart Professor of Medicine at Harvard Medical School and Physician-in-Chief at Beth Israel Hospital in Boston. Dr. Hiatt's major clinical and research interest has been the study of cancer.

Michito Ichimaru, M.D., is a specialist in leukemia and other blood diseases of atomic bomb survivors. He is Professor of Hematology at the Atomic Disease Institute of Nagasaki Medical School.

Leonid A. Ilyin, M.D., is the Director of the Institute of Biophysics of the USSR Ministry of Public Health, member of the Presidium of the USSR Academy of Medical Sciences, Chairman of the Committee on Radiologic Protection for the Soviet Union, and Vice Chairman of the Soviet Committee "Physicians for the Prevention of Nuclear War."

Penny Janeway, a freelance writer, is working on a project to educate medical professionals and the public about the medical consequences of nuclear weapons at the Center for the Analysis of Health Practices at the Harvard School of Public Health.

Patricia J. Lindop, M.D., is Professor of Radiation Biology at St. Bartholomew's Hospital at the Medical College of London. She is Chairman of the University of London Board of Studies in Radiation Biology.

Bernard Lown, M.D., is Professor of Cardiology at Harvard School of Public Health and a physician at the Brigham and Women's Hospital in Boston. Dr. Lown is co-founder and the first president (in the 1960's) of Physicians for Social Responsibility, and co-founder and President of International Physicians for the Prevention of Nuclear War.

Takeshi Ohkita, M.D., is Director of the Research Institute for Nuclear Medicine and Biology at Hiroshima University, where he is Professor of Hematology. He is one of the world's authorities on the medical effects of the bombings at Hiroshima and Nagasaki and is a major contributor to the book *Hiroshima and Nagasaki: The Physical, Medical and Social Effects of the Atomic Bombings* (1981).

Victor N. Petrov, Ph.D., is Head of the Department of Atmospheric Science at the Institute of Applied Geophysics in the USSR and an authority on atmospheric changes following nuclear explosions.

Joseph Rotblat, Ph.D., D.Sc., is Emeritus Professor of Physics at the University of London. He is one of the signers of the Russell–Einstein Manifesto and a former Secretary-General of Pugwash. He was also President of the British Institute of Radiology, Editor-in-Chief of *Physics in Medicine and Biology*, and chief physicist to St. Bartholomew's Hospital in London.

Naomi Shohno, Ph.D., is Professor of Theoretical Physics at Hiroshima Jogakuin College and Guest Professor at the Peace Research Center of Hiroshima University. He is also Chairman of the Hiroshima Society for the Study of Nuclear Problems and author of *Nuclear Radiation and Atomic-Bomb Disease* (1975).

Lewis Thomas, M.D., is Chancellor of the Memorial Sloan-Kettering Cancer Center and Professor of Pathology and Medicine at Cornell University Medical College. He is the former Dean of New York University's Bellevue Medical Center and of Yale University School of Medicine. In 1978 he received an award for medical journalism from the American Medical Writers' Association. He also received the National Book Award in Arts and Letters for his book *The Lives of a Cell* and the American Book Award in 1981 for *The Medusa and the Snail*.

Kosta Tsipis, Ph.D., is Director of the MIT Program in Science and Technology for International Security in the Physics Department of the Massachusetts Institute of Technology.

LAST AID

The Medical Dimensions
of Nuclear War

We are deluged with facts, but we have lost, or are losing, our human ability to *feel* them.

The real defense of freedom is imagination, that feeling-life of the mind which *actually* knows because it involves itself in its knowing, puts itself in the place where its thought goes . . . .

Archibald MacLeish
"A Country Journey"
Poetry and Journalism

Prologue

John E. Mack, M.D.

The nuclear arms race is out of control. Nuclear weapons of unimaginable destructiveness and great accuracy are being produced in ever greater numbers. The technology that governs their use is passing out of human hands. The probability of nuclear war is growing. An unlimited nuclear exchange—there is little likelihood that it would remain limited—would be a catastrophe of such proportions as to destroy civilization as we know it.

The production of nuclear weapons is rationalized on the grounds that they will prevent war. The capacity of a weapon to deter a potential enemy depends upon the belief that the other side would use the weapon in a conflict situation. But the actual use of nuclear weapons, in the words of General Dwight D. Eisenhower, would be "just another way of committing suicide." As more and more military leaders are realizing, such weapons serve no purpose other than to threaten mutual annihilation.

Why then do we continue to produce these suicidal devices, which have held the world hostage for nearly three decades? The answer lies not just in the technology of weapons production (where new scientific discoveries inevitably lead to the development of new weapons) and not just in the politics and economics of international relations. The ultimate responsibility for the arms race resides in the hearts and minds of human beings who are unable to comprehend the real nature of the monster we have created.

This book has been compiled in the hope of promoting understanding of nuclear weapons effects and of breaking through popular misconceptions about surviving nuclear war. These are first steps in freeing ourselves from the threat of self-destruction. We can regain our humanity only by moving from cold statistics and technical jargon to an emotional realization that through the continued production of these weapons (in such numbers and under the conditions of tension that exist in the world), the nuclear nations are risking the extermination of helpless human beings. Only by understanding the agonizing suffering and slow death of countless others through burning, crushing, and radiation sickness can we wrest the control of human destiny from the destructive instruments we have so heedlessly spawned. The projected "kill-ratios," the sterile "defensive" acronyms, the "tactics and strategies" must be transformed into a realistic awareness of the imminent danger we face.

Three of the four editors of this volume are psychiatrists. This is not accidental. Students of nuclear arms are recognizing that serious emotional consequences derive from living with the threat of nuclear war and that the perpetuation of the arms race is in part the result of pathological distortions of thinking. The need for a better understanding of psychological avoidance patterns has become pressing. These patterns, which contribute to an individual's sense of well-being in the short run, may in the long term represent a grave danger to the survival of our species.

It is not hard to see why we find the danger of nuclear weapons so difficult to grasp. First there is the problem of scale. We wince at the thought of a gas explosion next door, which might cause severe burns to our neighbors, or fractures and bleeding injuries to their children. But who can grasp the reality that the United States and the Soviet Union have aimed at each other approximately 50,000 nuclear warheads, any one of which could burn, maim, or kill hundreds of thousands of people, or that a single nuclear submarine is fitted with enough missiles to destroy most of the major cities in both our countries?

A second difficulty is the natural tendency of the human mind to turn away from a thought that is too terrifying to contemplate, especially if the dangerous event can be made to seem remote. The "unreality" of nuclear weapons can create thought distortion,

> We go on piling weapon upon weapon, missile upon missile, new levels of destructiveness upon old ones ... helplessly, almost involuntarily: like the victims of some sort of hypnosis, like men in a dream, like lemmings heading for the sea, like the children of Hamelin marching blindly behind their Pied Piper.
>
> George Kennan
> Former U.S. Ambassador to the USSR
> May 19, 1981

such as the illusions of escape that underlie civil defense planning. The following statement by a former Army General is an example of make-believe about our capacity to survive a nuclear attack: "All you need to do to escape the effects of a one-megaton bomb is to walk for an hour and hide behind a bush."

A third obstacle to grasping the reality of the nuclear threat is the failure to distinguish between intellectual and emotional knowing. Most of us do not experience the reality of nuclear war emotionally. Children and adolescents seem better able to look squarely at the actuality of the nuclear danger than adults and to recognize the potential impact of this threat on their lives. Perhaps this is because their psychological defenses are less formed. Also, because of their youth, they have a greater stake in the future. Recent studies suggest that the threat of nuclear annihilation is a dominating fear for many children and adolescents. A 17-year-old boy, asked by a *Boston Globe* interviewer how old he was when he first became aware of nuclear issues, responded: "Very young, seven or eight. It was in a dream. I didn't know what the dream was at the time. I first felt intense fear and complete and utter destruction. This dream came back throughout my childhood and it wasn't until five or six years ago that I figured out that this dream was a nuclear holocaust. Thinking of this scares me more than anything I have yet known."

There is a profound uncertainty about the future, resentment toward adults for creating the situation, and disillusionment with political leaders. They perceive a world out of control, and a nuclear holocaust likely in the next decade if there is not change in the direction we have taken.

Another eleventh grader explained his feelings about living with the fear that the world might not be around to grow up in. "No one should have the right to choose whether an entire generation gets to grow up or not. What could possibly be more simple? No one has that right. This has to be made clear."

Most adults avoid knowing in the full emotional sense how close the world is to the use of nuclear weapons and the subsequent havoc that would lay waste our society. There are, probably, some few individuals who actively avoid this awareness out of a political or personal stake in weapons production and the arms race. But most of us—and this applies to political and military leaders as well—turn away from acknowledging the reality of the nuclear danger out of fear.

The experience of the last few years suggests that when men and women, whatever their role in society, can allow themselves truly to contemplate the enormity of the nuclear danger, they begin to commit themselves to the process of reducing the threat of nuclear annihilation. They become responsibly involved. One 13-year-old boy, after taking a course in his school called "Decision-Making in a Nuclear Age," said that had he not taken the course he would not be so scared. But, he continued, "That's reality," and "Because I took it and I know nuclear war is a possibility, I'll try to do something to stop it."

We witness at this moment in history a growing moral revulsion against nuclear weapons, led by responsible citizens, including clergymen and doctors. Physicians in all countries are committed to the preservation of life. The extent of death and suffering that a nuclear war would cause and the knowledge that there is no effective medical response once an attack has taken place are forcing doctors all over the world to become active in efforts to prevent such a war. Together we can, *we must*, rid humankind of this intolerable danger.

SECTION I

Introduction

Any nuclear war would inevitably cause death, disease and suffering of pandemic proportions and without the possibility of effective medical intervention. The only hope for humanity is prevention of any form of nuclear war.

Pope John Paul II
Vatican City
January 1, 1982

1

Physicians and Nuclear War

Bernard Lown, M.D.

The world is moving inexorably toward the use of nuclear weapons. The atomic age and space flight demonstrate the awesome power of science and technology. These developments have also brought humankind to a bifurcation—one road of unlimited opportunity for improving the quality of life, the other of unmitigated misery, devastation, and death. The road we follow will determine whether modern society has a future.

In Hiroshima, a primitive uranium bomb with an explosive power of 13 kilotons (equivalent to 13,000 tons of TNT) instantaneously killed more than 75,000 persons. Of a population of 245,000, more than 100,000 persons were injured. Ninety percent of the 76,000 buildings within city limits were destroyed. Only 3 of Hiroshima's 45 hospitals were left unscathed. Fewer than 30 of the 150 physicians were available to attend the thousands of victims. Of the 1780 nurses, only 126 were alive and able to assist the burned and traumatized casualties. Long after the bombing, illness and affliction continued to emerge. Some of the people living in proximity to the nuclear detonation experienced radiation

From the *Journal of the American Medical Association,* 246 (no. 20), November 20, 1981. Copyright 1981, American Medical Association. The illustrations have been added.

Hiroshima panorama, August 12, 1945. The view is from the roof of the Higashi Police Station.

effects years to decades later, including leukemia and cancer of the breast, lung, and thyroid.

With the passage of years, the bomb appeared less threatening. After all, both Hiroshima and Nagasaki continued as vibrant cities. Within hours after the bombing, aid arrived from other parts of Japan. Many of the victims were evacuated to adjacent communities with largely intact medical facilities. Both cities have been rapidly rebuilt. The aftermath of the bombing is now only to be recognized in museums, in commemorative monuments, and in surviving victims. In effect, the consequences of nuclear bombs, contrary to numerous prophecies of doom, could be overcome. But there are a number of new realities that need examination.

The United States and the Union of Soviet Socialist Republics, the two superpowers, now possess more than 40,000 nuclear weapons for a total of about 15,000 megatons. Even a single megaton carries unimaginable destructive power, more than 70 times that which destroyed Hiroshima. A megaton is equivalent to one million tons of TNT, and were this to be transported by rail, the train would stretch more than 640 kilometers (400 miles). If one were to add the weapons in the arsenals of the other nuclear powers, the 20,000-megaton total, now in world stockpiles, pro-

vides 5 tons of explosive power for every man, woman, and child on this earth.

The Hiroshima and Nagasaki experiences, horrendous as they were, have but limited validity in extrapolating the consequences of nuclear war with present-day arsenals. Kiloton-sized weapons have been multiplied a thousandfold; instead of individual cities, there are now multiple-city targets. No help could be forthcoming from adjoining or distant communities, which likewise could be victims of blast, firestorm, and radiation. In one likely and specific scenario, during an all-out nuclear war in mid-1980 more than 200 million men, women, and children would be killed outright. More than 60 million would be seriously injured; among these, half would experience radiation sickness. The remainder would be afflicted with various combinations of trauma, burns, and irradiation. Crushing injuries, fractures, perforated viscera, contusions of internal organs, charred extremities, and uncertain radiation exposures, as well as psychiatric derangements, would be commonplace. Such injuries would tax the best of health care systems during peacetime conditions. Even a one-megaton explosion above a city of one million people would produce at least ten times more severely burned patients than could be accommodated by all burn treatment facilities in the United States. Given disrup-

9

[The Soviets] don't want nuclear war any more than we do. . . . They lost 20 million people. They talk to you about it all the time. They don't want to see their children go through the hell that they went through.

Harold Brown
Former U.S. Secretary of Defense
"Meet the Press," March 1977

tion of the health care system, its lack of transport, communication facilities, electric power, hospital beds, x-ray and other diagnostic equipment, blood, drugs, and essential instruments, surviving health workers could provide only verbal comfort to those in pain and dying. As Khrushchev direly prophesied, "The living will envy the dead."

Yet even these extrapolations do not encompass many potential long-range effects. They exclude possible natural consequences, such as degradation of the stratospheric ozone layer, long-term climatic changes, radioisotope contamination of the food chain, crop failures resulting from destruction and insect ecology, and yet other predictable adverse consequences.

A superficial analysis would suggest that society and the nuclear bomb have reached a tolerable symbiosis. The fact that these weapons have been around for 35 years and have not been used by the superpowers against one another promotes wide public belief of the permanence of such stability. In short, we have learned to live with the bomb. This is a dangerous illusion. In the first place, there is the escalating nuclear arms race. The destructive power stored in the world arsenals now exceeds by a factor of more than a million the bomb that destroyed Hiroshima. The multiplication of nuclear weapons is intended to promote a nation's security, but the cost is that of increasing world insecurity. History provides scant comfort for the view that peace is promoted by the preparation for war. The arms race cannot be a process without end; its terminus is inevitable nuclear confrontation.

More worrisome than the growth in sheer numbers of weapons or their proliferation is the qualitative transformation in the dy-

American war deaths

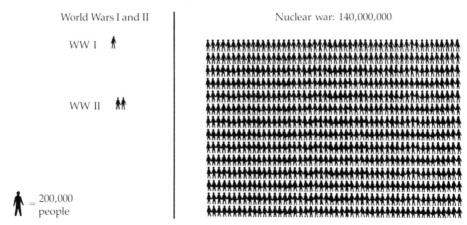

World Wars I and II

WW I

WW II

$\mathbf{\dot{\intercal}}$ = 200,000 people

Nuclear war: 140,000,000

Soviet war deaths

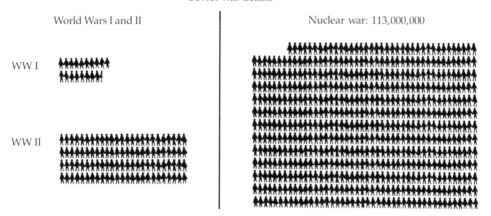

World Wars I and II

WW I

WW II

Nuclear war: 113,000,000

American and Soviet war deaths. (Charts by Arthur Kanegis, Center for Defense Information, Washington, D.C.; nuclear war estimates by the Military Strategy and Force Posture Review, Presidential Review Memorandum #10.)

namics of the atomic arms race itself. Until recently, an uneasy peace has prevailed, assured by the doctrine of deterrence or the certainty of incalculable damage to one's adversary. Each side has acknowledged that protection of its own citizenry is based on threat of annihilation of the opposing side. In a sense, nearly 500

11

million people have been held hostage by the two superpowers. This policy of mutual assured destruction, the acronym for which is MAD, is now undergoing alteration. Changes in strategic doctrine are largely propelled by technological innovation. Development of great precision in so-called terminal guidance systems has increased accuracy of targeting. This makes missiles, in their hardened silos, vulnerable. These weapons then lose value in retaliation. A symmetrical instability ensues wherein the temptation for an enemy to attack is at the same time inducement for one's own preemption. These developments in turn create great uncertainty and international instability.

The instability is enhanced by the time factor—in a period of crisis, little time is available for making an awesome decision. Only 30 minutes would elapse between launch and impact of nuclear weapons between the USA and the USSR. This imposes dependence on complex technologies to monitor, detect, and analyze signals of attack. The increased accuracy of guidance systems invites contemplation of preemptive strikes, promotes "hair-trigger" readiness, and increasingly compels reliance on computers to sort out the real from spurious warnings of a potential attack. Physicians in their daily work know the malfunctioning of technological hardware, but while the failure of a pacemaker, a defibrillator, or an oxygenator might jeopardize a single life, the malfunctioning of military technology threatens a way of life. We have set in motion forces over which we shall increasingly have less control. Ultimately, the bomb may take command.

Moreover, the arms race is imposing enormous economic, psychological, and moral costs. In the 1970's, world military expenditures exceeded $4 trillion ($4,000,000,000,000), which is greater than the total goods and services created by humankind in one year. Forty percent of research budgets are devoted to the military. Last year, more than $500,000 million was spent by the world's military establishment. This is equivalent to $1400 million per day, or $1 million per minute. This massive diversion of scarce resources diminishes the development of knowledge, technology, and human resources that could address global ecological and overpopulation problems. A small fraction of these military expenditures could provide the world with adequate food and sanitary water supply, housing, education, and modern health care.

Physicians can cite numerous illustrations of the power of modest investments in health. For example, slightly more than a decade ago, smallpox was endemic in 33 countries with a total population of 1200 million, with 10 to 15 million cases and loss of 2 million lives annually. The World Health Organization campaign for smallpox eradication was successfully completed in a decade. The last case of naturally occurring smallpox was diagnosed in Somalia in 1977. This single achievement required an investment of $300 million, approximately five hours of the cost of military budgets. With a diversion of funds consumed by three weeks of the arms race, the world could obtain a sanitary water supply for all of its inhabitants. The lack of clean water now accounts for 80 percent of all the world's illness and imposes untold misery and degradation on half the world's population.

In addition to economic costs, the nuclear arms race has incurred psychological costs as well. Living with the possibility of imminent annihilation has created a new reality for humanity with profound and widespread psychological effects. Not only does each person have to deal with the possibility of his or her own agony or sudden death, but also with the dissolution of the total surroundings. Membership in a family, community, society, and nation provides the psychological means for coming to terms with individual death. Something of the "I" survives in the social germ plasm, providing a symbolic continuity or immortality. But suppose nothing survives! How does a person come to terms with nothing? And what are the psychological injuries in confronting a world where extinction can be so absolute?

Little is said of the moral degradation deriving from the preparation for nuclear war. Bertrand Russell called attention to this aspect some 30 years ago: "Our world has sprouted a weird conception of security and a warped sense of morality—weapons are sheltered like treasures, while children are exposed to incineration."

Physicists who let the atomic genie out of the bottle did not succeed in either stemming the nuclear arms race or persuading humankind of the ominous peril to their lives. We physicians must join our scientific colleagues who have insistently, with an increasing sense of despair and urgency, attempted to alert humanity. Only an aroused world public opinion can compel political leaders to stop the spiraling nuclear arms race.

Physicians, as few other groups in society, are committed to maintaining health and promoting survival. We have a world

*A child suffering from cholera. Polluted water is poured over his forehead to cool the fever. India, 1971. (Donald McCullin/*The Sunday Times, *London.)*

constituency of four million members who have taken a sacred and ancient oath to assuage human misery and preserve life. In the face of the nuclear threat, this commitment imposes social and moral obligations for us to band together, to make our collective voices heard and, it is hoped, heeded. If we are to succeed, we must realize our own limitations. We can have credibility and be effective only as long as we scrupulously adhere to the province of our expertise as scientists and as healers. We must not become bogged down in debating the political differences that have fueled the cold war and have hindered détente. We are not politicians. Nor are we arms control experts; we cannot discourse or debate over weapons systems, deterrence, retaliation, or overkill. We can speak on the threat of nuclear weapons, on the consequences of nuclear war, on the diversion of scarce resources from human needs, on the psychological, moral, and biological implications of the arms race. We must recognize the one fundamental reality of this nuclear age, that the futures—the very fate—of American, European, and Soviet societies are indissolubly linked. We shall either live together or die together.

We are but transient passengers on this planet Earth. It does not belong to us. We are not free to doom generations yet unborn. We are not at liberty to erase humanity's past or dim its future. Social systems do not endure for an eternity. Only life can lay claim to uninterrupted continuity. This continuity is sacred. We physicians, who shepherd human life from birth to death, are aware of the resiliency, courage, and creativity that human beings possess. We have an abiding faith in the concept that humanity can control what humanity creates. This perception provides optimistic purpose in reversing the direction of humankind's potential tragic destiny.

2

Physicians for Nuclear Disarmament

Evgueni I. Chazov, M.D.

Life on earth has never been in such danger as now. This danger is because of the nuclear arms race and the production of weapons whose devastating power cannot be compared to that of any earlier types of weaponry. Europeans recall with horror World War II, which snuffed out 50 million lives, leaving behind ruined cities and tragedies for millions of families. Twenty million people were killed in the USSR. Virtually every family was affected.

We know very well what war means. But what could happen to our planet in no way compares to what humanity lived through during World War II. It has been calculated that about five megatons of various kinds of explosive substances were used during the whole of that war. To get an idea of what could happen today as a result of nuclear war, we must remember that the explosive power of just one thermonuclear charge is several times greater than the total of all explosions made in the course of all wars.

The radioactivity from nuclear weapons is as devastating to human life as is their destructive power. Even an hour after a one-megaton nuclear explosion, the radioactivity at the place of explosion is equal to the radioactivity of 500 million kilograms of radium. It is scores of millions of times greater than that which is included in the powerful gamma installations used in medicine for the treatment of malignant tumors. Few people in the world

Soviets searching for their dead, World War II. (Dmitri Baltermants, Sovfoto.)

American war dead, D-Day. Omaha Beach, Normandy, June 6, 1944.
(Robert Capa from Magnum for Life.*)*

17

> **Many experts, physicians, and physicists realistically assess**
> **this danger (nuclear war). They understand that one cannot**
> **wait idly as the threat to the very life of mankind hangs over**
> **every day of existence.** *But the broad masses, of course, have*
> *not fully realized all the horrors such a war can bring. And*
> *the importance of our movement is especially to bring the*
> *truth of the catastrophic consequences of nuclear conflict to*
> *the public.*
>
> E. I. Chazov
> *Komosol Pravda* interview
> April 10, 1981

realize the dire consequences a nuclear war would have for each
one of us, for our loved ones, for humanity. The total explosive
power of the nuclear arsenals stockpiled in the world today is
equal to a million bombs of the kind dropped on Hiroshima. Figu-
ratively speaking, we are now sitting on a powder keg that holds
about 5 tons of TNT for each one of us. Around this powder keg,
certain people are waving the torch of a "nuclear policy strategy,"
which at any second, even accidentally, may cause an explosion
that would be a world catastrophe.

Little by little, with the joys and sorrows of daily life, people
forget the horrors of Hiroshima and Nagasaki. Some military and
public functionaries and even some scientists are trying to mini-
mize the possible consequences of nuclear war. Statements appear
that a nuclear war can be won; that a limited nuclear war can be
waged; that humanity and the biosphere will persist, even in conditions of total nuclear catastrophe.

These are illusions that must be dispelled. Hiroshima and
Nagasaki are a reality, historical facts that show that science and
technology have fostered power that can lead to global annihilation of every living thing. The bell that tolls in Hiroshima re-
minds people of the danger. It calls on them to be vigilant. It calls
on them to do their utmost to prevent the tragedy of nuclear war.

Even now the nuclear arms race is very costly for humanity.
Serious psychological harm stems from the fear experienced by
people all over the world as a result of the threat of a nuclear war.

European war deaths
(not counting USSR)

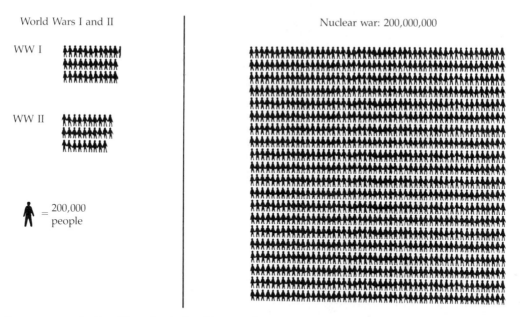

| World Wars I and II | Nuclear war: 200,000,000 |

WW I

WW II

= 200,000 people

European war deaths. (Chart by Arthur Kanegis, Center for Defense Information, Washington, D.C.; nuclear war estimate by Henry Kendall, Professor of Physics, MIT.)

Tremendous sums are being spent on the technology of a nuclear war at a time when millions of people go hungry, when they suffer from disease, when illiteracy is still widespread. Huge expenditures on manpower and material resources render difficult the solution of numerous world problems—health, energy, education, economics, and more.

In developing countries today, 100 million children are in danger of dying because of malnutrition and vitamin shortage, and 30 percent of the children have no possibility of going to school. Yet military expenditures worldwide are 20 to 25 times bigger than the total aid provided annually to the developing countries by the developed states. For example, in the past ten years the World Health Organization spent about $83 million on smallpox eradication—less than the cost of one modern strategic bomber. According to some World Health Organization calculations, $450 million is needed to eradicate malaria, a disease that affects more than

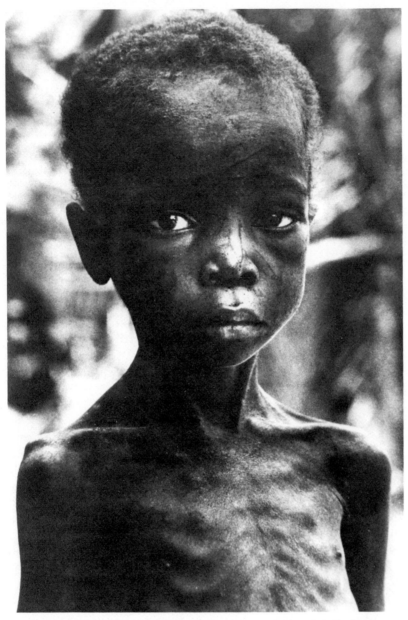

Nine-year-old Biafran boy waiting for food, 1970. (Donald McCullin/The Sunday Times, *London.)*

We understand that we only took the first step. But at the beginning there were only six of us, then thirty-six, then one hundred and fifty. Now millions of people in the world know about our ideas, and I hope that our movement [International Physicians for the Prevention of Nuclear War] will gain strength and will begin to exert influence on political leaders so that a nuclear war, which will have no winners, can be prevented. There is no alternative.

E. I. Chazov
Komosol Pravda interview
April 10, 1981

1000 million people in 66 countries of the world. This is less than half of what is spent in the world on arms every day.

Outstanding scientists and physicians have realized the threat to humanity of the nuclear arms race and understood the need to work for a ban on nuclear weapons. Twenty-seven years ago Albert Einstein, Bertrand Russell, Frederic Joliet–Curie, Joseph Rotblat, and others issued the Russell–Einstein Manifesto, which began the Pugwash movement. They stated: "In the tragic situation which confronts humanity, we feel that scientists should assemble in conference to appraise the perils that have arisen as a result of the development of weapons of mass destruction." And they warned: "We are speaking on this occasion, not as members of this or that nation, continent, or creed, but as human beings, members of the species Man, whose continued existence is in doubt."

Since 1957 leading scientists from around the world have held regular Pugwash meetings, focusing their expertise on the dangers of nuclear weapons and urging governments to eliminate these weapons from their arsenals. Yet the world has not heeded the warnings of these great scientists.

Today, the four million physicians working all over the world must regard the struggle against the danger of a nuclear war to be not only the duty of an honest, humanitarian person, but also a professional duty. We are dealing not with political problems, but with the preservation of the health and lives of all people. No country or people will remain unaffected by a nuclear catastrophe.

21

Raising our voice of protest against the arms race and nuclear war, we must at the same time find the most effective path for our cause. We must explain to the peoples of the world and to governments possessing nuclear weapons, on the basis of our knowledge and precise research data, the danger to life on earth from the unleashing of nuclear war. We must discuss not only the immediate consequences of a nuclear explosion, but also the global problems resulting from the radioactive contamination of the stratosphere—the disruption of the ozone layer of the earth, the changes of the climate, ecology, and more.

We must convince the peoples and the governments that under conditions of nuclear war, medicine will be unable to provide aid to the hundreds of thousands of wounded, burned, and sick, because of the deaths of doctors and the destruction of the transportation system, drugs, blood supplies, hospitals, and laboratories. Epidemic outbreaks will reach far beyond the affected centers.

As physicians, our knowledge of the tragic consequences of nuclear war enables us to make a vital contribution to the cause of preventing it. Our patients entrust themselves to us. It is in keeping with our professional honor and with the oath of Hippocrates that we have no right to hide from them the danger that now threatens us all.

We must preserve life on earth. We must struggle for the survival of our children and grandchildren. People of all political outlooks, nationalities, and religions must urge their governments to concentrate their attention not on what steps to take to attain victory in nuclear war, but on what must be done so that the flames of such a war will never burn on our planet.

We face many difficulties, and the path to the achievement of our goal will be thorny and full of impediments. But we have no alternative: when we hear that "humanity is in danger," physicians must rise to the challenge. We call on all the physicians of the world, on all medical workers, to merge their efforts for the salvation of life on earth.

3

General Principles
of Nuclear Explosions

Edited by Samuel Glasstone, Ph.D.,
and Philip J. Dolan, Ph.D.

An explosion, in general, results from the very rapid release of a
large amount of energy within a limited space. This is true for a
conventional "high explosive," such as TNT, as well as for a nuclear (or atomic) explosion, although the energy is produced in
quite different ways. The sudden liberation of energy causes a
considerable increase of temperature and pressure, so that all the
materials present are converted into hot, compressed gases. Since
these gases are at very high temperatures and pressures, they expand rapidly and thus initiate a pressure wave, called a "shock
wave," in the surrounding medium—air, water, or earth. The
characteristic of a shock wave is that there is (ideally) a sudden
increase of pressure at the front, with a gradual decrease behind
it. A shock wave in air is generally referred to as a "blast wave"
because it resembles and is accompanied by a very strong wind. In
water or in the ground, however, the term "shock" is used, because the effect is like that of a sudden impact.

Nuclear weapons are similar to those of more conventional
types insofar as their destructive action is due mainly to blast or

Reprinted from *The Effects of Nuclear Weapons,* 3rd ed., compiled and edited by
Samuel Glasstone and Philip J. Dolan, U.S. Department of Defense and U.S. Department of Energy, 1977.

shock. On the other hand, there are several basic differences between nuclear and high-explosive weapons. In the first place, nuclear explosions can be many thousands (or millions) of times more powerful than the largest conventional detonations. Second, for the release of a given amount of energy, the mass of a nuclear explosive would be much less than that of a conventional high explosive. Consequently, in the former case, there is a much smaller amount of material available in the weapon itself that is converted into the hot, compressed gases mentioned above. This results in somewhat different mechanisms for the initiation of the blast wave. Third, the temperatures reached in a nuclear explosion are very much higher than in a conventional explosion, and a fairly large proportion of the energy in a nuclear explosion is emitted in the form of light and heat, generally referred to as "thermal radiation." This is capable of causing skin burns and of starting fires at considerable distances. Fourth, the nuclear explosion is accompanied by highly penetrating and harmful invisible rays, called the "initial nuclear radiation." Finally, the substances remaining after a nuclear explosion are radioactive, emitting similar radiations over an extended period of time. This is known as the "residual nuclear radiation" or "residual radioactivity."

It is because of these fundamental differences between a nuclear and a conventional explosion, including the tremendously greater power of the former, that the effects of nuclear weapons require special consideration. In this connection, a knowledge and understanding of the mechanical and the various radiation phenomena associated with a nuclear explosion are of vital importance.

All substances are made up from one or more of about 90 different kinds of simple materials known as "elements." Among the common elements are the gases hydrogen, oxygen, and nitrogen; the solid nonmetals carbon, sulfur, and phosphorus; and various metals, such as iron, copper, and zinc. A less familiar element, which has attained prominence in recent years because of its use as a source of nuclear energy, is uranium, normally a solid metal.

The smallest part of any element that can exist, while still retaining the characteristics of the element, is called an "atom" of that element. Thus, there are atoms of hydrogen, of iron, of uranium, and so on, for all the elements. The hydrogen atom is the lightest of all atoms, whereas the atoms of uranium are the heaviest of those found on earth. Heavier atoms, such as those of pluto-

Shallow underwater nuclear explosion dwarfs full-size naval vessels. The circular shock wave can be seen on the water's surface. Bikini Atoll, 1946. (U.S. Air Force.)

This basic power of the universe cannot be fitted into the outmoded concept of narrow nationalisms. For there is no secret and there is no defense, there is no possibility of control except through the aroused understanding and insistence of the peoples of the world.

Albert Einstein
Letter as Chairman, Emergency Committee of Atomic Scientists
January 22, 1947

nium, also important for the release of nuclear energy, have been made artificially. Frequently, two or more atoms of the same or of different elements join together to form a "molecule."

Every atom consists of a relatively heavy central region or "nucleus," surrounded by a number of very light particles known as "electrons." Further, the atomic nucleus is itself made up of a definite number of fundamental particles, referred to as protons and neutrons. These two particles have almost the same mass, but they differ in that the proton carries a unit charge of positive electricity whereas the neutron, as its name implies, is uncharged electrically, i.e., it is neutral. Because of the protons present in the nucleus, the latter has a positive electrical charge, but in the normal atom this is exactly balanced by the negative charge carried by the electrons surrounding the nucleus.

The essential difference between atoms of different elements lies in the number of protons (or positive charges) in the nucleus; this is called the "atomic number" of the element. Hydrogen atoms, for example, contain only 1 proton, helium atoms have 2 protons, uranium atoms have 92 protons, and plutonium atoms 94 protons. Although all the nuclei of a given element contain the same number of protons, they may have different numbers of neutrons. The resulting atomic species, which have identical atomic numbers but which differ in their masses, are called "isotopes" of the particular element. All but about 20 of the elements occur in nature in two or more isotopic forms, and many other isotopes, which are unstable, i.e., radioactive, have been obtained in various ways.

Each isotope of a given element is identified by its "mass number," which is the sum of the numbers of protons and neutrons in the nucleus. For example, the element uranium, as found in nature, consists mainly of two isotopes with mass numbers of 235 and 238; they are consequently referred to as uranium-235 and uranium-238, respectively. The nuclei of both isotopes contain 92 protons—as do the nuclei of all uranium isotopes—but the former have in addition 143 neutrons and the latter 146 neutrons. The general term "nuclide" is used to describe any atomic species distinguished by the composition of its nucleus, i.e., by the number of protons and the number of neutrons. Isotopes of a given element are nuclides having the same number of protons but different numbers of neutrons in their nuclei.

In a conventional explosion, the energy released arises from chemical reactions; these involve a rearrangement among the atoms, e.g., of hydrogen, carbon, oxygen, and nitrogen, present in the chemical high-explosive material. In a nuclear explosion, on the other hand, the energy is produced as a result of the formation of different atomic nuclei by the redistribution of the protons and neutrons within the interacting nuclei. What is sometimes referred to as atomic energy is thus actually nuclear energy, since it results from the particular nuclear interactions. It is for the same reason, too, that atomic weapons are preferably called "nuclear weapons." The forces between the protons and neutrons within atomic nuclei are tremendously greater than those between the atoms; consequently, nuclear energy is of a much higher order of magnitude than conventional (or chemical) energy when equal masses are considered.

4

The Physical Effects
of a Nuclear Explosion

Kosta Tsipis, Ph.D.

When a nuclear weapon is exploded, a very large amount of energy is released within a very short period of time. Where does this energy come from? In what forms is it released? What are the effects of this release? Each of these questions will be addressed in the following pages.

Where Does This Energy Come From?

The protons and neutrons in the nucleus of an atom are called nucleons. The sum of the number of nucleons is called the mass number or atomic weight. It is an experimental fact that the actual weight of the nucleus (m_N) is less than the sum of the actual weights of the protons (m_p) plus the neutrons (m_n). Because protons carry positive charges, they repél each other, so that energy is needed to bind the nucleus together to overcome the repulsion.

According to Einstein's formula $E = mc^2$, this binding energy (E_B) is equal to the difference between the weight of the nucleus and the sum of the weights of its constituents times the velocity of light squared (c^2):

$$E_B = [m_N - (m_p + m_n)] \times c^2$$

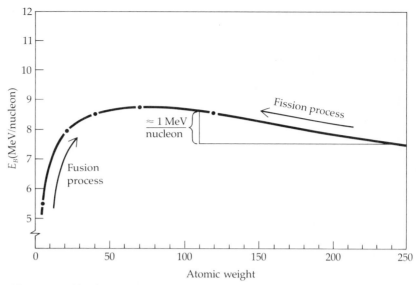

The curve of binding energy.

Plotting the binding energy per nucleon against the atomic weight yields the experimental curve called the curve of binding energy (see the figure above).

When a nucleus of uranium-235 fissions (splits) into two pieces, the binding energy per nucleon in the daughter nuclei is higher than the binding energy each nucleon of the parent nucleus is subjected to. Since each nucleon in each daughter nucleus is more tightly bound, it means that it can move less freely; the nucleons have lost some of their kinetic energy. The amount of energy each nucleon has lost is equal to the amount by which each nucleon is more tightly bound after the fission. From the figure above, we see that this amount is about one million electron volts (1 MeV) per nucleon. Since this kinetic energy cannot simply disappear, it is released into the world outside the fissioned nucleus. Uranium-235 has 235 nucleons; therefore, the total energy released by a single fission of a uranium nucleus is, on the average, about 200 MeV.

When a plutonium (atomic weight = 239) or uranium-235 nucleus is hit by a speeding neutron, it fissions, releasing energy and producing, on the average, two additional neutrons energetic enough to split other plutonium-239 or uranium-235 nuclei. These

29

in turn produce four additional neutrons; there is thus a doubling
of neutrons available for fissioning nuclei with each generation.

This doubling is the basis for a chain reaction, which releases
the enormous energy for both nuclear power plants and nuclear
weapons. There are 6×10^{23} nuclei in 235 grams (about one-half
pound) of uranium-235, so to fission all these nuclei completely,
we need 6×10^{23} neutrons. If we start with one neutron, we will
need 79 doubling steps to produce that many neutrons. These
neutrons travel at almost the speed of light; to move from one
uranium-235 nucleus to another takes one–ten-thousandth of a
millionth of a second (0.000 000 000 1 second). The total time,
then, necessary to fission all the uranium-235 nuclei is about 80
times that or eight-thousandths of a millionth of a second
(0.000 000 008 second).

In What Forms Is the Energy Released?

The nuclear weapon used on Hiroshima contained uranium-235;
that used on Nagasaki contained plutonium. If we examine the
fissioning of 10 kilograms of plutonium, a sufficient amount to
make such a weapon, we can learn how to calculate the energy
released by an atomic weapon. A sphere of 10 kilograms of
plutonium-239 has a diameter of about 20 centimeters (about
8 inches—the size of a cantaloupe) and contains about 2.5×10^{25}
nuclei. If all were fissioned, the energy released would be $6 \times
10^{27}$ MeV or 6×10^{14} joules. However, only a few percent of the
material in a nuclear weapon would fission, so the actual energy
released would be about 5×10^{13} joules. The temperature generated when this much energy is released in this volume would be

30

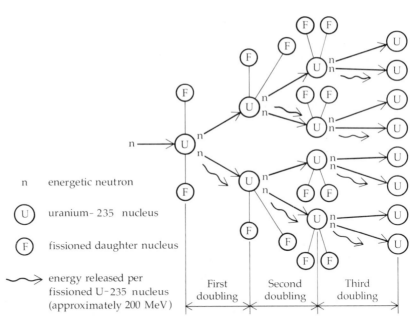

n   energetic neutron

(U)   uranium-235 nucleus

(F)   fissioned daughter nucleus

⇝   energy released per
fissioned U-235 nucleus
(approximately 200 MeV)

First doubling	Second doubling	Third doubling

The chain reaction principle.

about 10^8 or 100 million degrees Celsius, and the accompanying pressure would be 10^8 or 100 million atmospheres (1 atmosphere pressure = 14.7 pounds per square inch). These are the conditions at the center of the sun. In fact, an exploding nuclear weapon is a miniature, instantaneous sun.

A thermonuclear weapon, also called a Hydrogen Bomb, is a bit more complex than that. In addition to the fissionable material, plutonium-239, a thermonuclear weapon contains deuterium (or heavy hydrogen, having one proton and one neutron in the nucleus) and tritium (with one proton and two neutrons). The heat from the fission process results in a fusion of deuterium and tritium (hence the name thermonuclear) to form helium (with an atomic weight of 4: two protons and two neutrons). This fusion process also releases large amounts of energy. (In the figure on p. 29, the slope is steeper for atomic weights below 20, so that the energy release for the fusion reaction is larger per nucleon than for the fissioning of uranium or plutonium.) Fusion of deuterium and tritium also releases a highly energetic neutron that can fission another form of uranium, uranium-238, which is not fissionable by a chain reaction.

31

Fireball from an airburst in the megaton energy range (photographed at an altitude of 12,000 feet from a distance of 50 miles). Bikini Atoll, 1956. (U.S. Air Force.)

A thermonuclear weapon works in this way. The plutonium is first squeezed by means of properly shaped high explosives into a volume called a critical mass at which it can sustain a chain reaction. The energy released by the fissioning of the plutonium causes the deuterium and tritium to fuse and release additional large amounts of energy plus a very large number of energetic neutrons. These neutrons in turn fission the uranium-238 blanket around the deuterium and tritium mixture. This process releases as much energy as the fusion of deuterium and tritium. Thus, a thermonuclear weapon is a fission-fusion-fission bomb, where in general 50 percent of the energy is provided by the fission and 50 percent by the fusion.

A Neutron (or Enhanced-Radiation) Bomb is also a thermonuclear weapon, but without the uranium-238 blanket. Without the blanket, there is a large excess of high-energy neutrons released in addition to the thermal blast and initial radiation released by the fission-fusion core.

The fission-fusion-fission process in a standard nuclear weapon results in an intensely hot, furiously expanding mass of radioactive nuclear fragments called the fireball, which for a one-megaton explosion is one-quarter to one-half mile in diameter. The fireball expands rapidly by two mechanisms:

- It emits x-rays that heat the immediately surrounding air, which then becomes transparent to the x-rays. More layers are thus exposed farther from the fireball, and these emit visible light.

- The incredible pressure inside the fireball compresses the surrounding air suddenly, shock-heating it until it becomes luminous. After the first second following a nuclear explosion, a glowing, superheated wall of air and a giant pulse of radiation (that changes from x-rays to visible light to thermal infrared rays) sweep away from the point of detonation, engulfing everything in its path.

What Are the Effects of This Sudden Release of Energy?

The expanding fireball creates a number of physical phenomena:
    Thermal radiation
    Prompt nuclear radiation
    Air-blast mechanical effects
      overpressure
      dynamic pressure
    Cratering
    Ground shock
    Electromagnetic pulse, induced currents
    Radioactivity
    Ozone depletion

These will be described for a nuclear weapon with an explosive energy equivalent to one million tons of TNT; warheads of this size largely constitute both U.S. and USSR arsenals. One million tons of TNT would fill a freight train 300 miles long and would produce enough energy to transform one million tons of ice into steam (4×10^{15} joules).

For a one-megaton surface nuclear explosion, the thermal radiation is capable of igniting most flammable materials at approximately $5\frac{1}{2}$ miles from the hypocenter. People with exposed skin will sustain third-degree burns up to a distance of approximately 7 miles and second-degree burns at approximately 8 miles from the hypocenter.

The ignition of multiple flammable materials in cities under nuclear attack could coalesce into a firestorm, which is a massive self-fanning fire that can cover 100 square miles. Firestorms can generate temperatures up to 2000 degrees Fahrenheit (enough to melt glass and sheet metal) and can produce massive amounts of carbon monoxide and carbon dioxide, which would asphyxiate the occupants of shelters in the areas. A firestorm can burn for days.

The explosion of a standard fission-fusion-fission weapon also generates prompt nuclear radiation. For a one-megaton burst, 1000 rads of high-energy neutrons reach a range of about 2000 yards from the hypocenter and 1000 rads of gamma radiation reach about 2700 yards, both of which are lethal doses for 100 percent of exposed healthy adults. Radiation zones for neutrons and gamma rays fall within the larger lethal zones of burn and blast effects.

A nuclear explosion in the air or on the ground creates a shock front of very high pressure that propagates outward from the point of detonation, crushing and sweeping everything in its path. The overpressure, caused by the instantaneous compression

---

Editors' note: Allied incendiary bombs exploded on hundreds of targets in Hamburg on the night of July 27, 1943. These fires merged, causing columns of intensely hot air to rise, with cool air rushing in to fill the vacuum. Winds of 150 miles per hour resulted, uprooting trees and flinging cars about. Temperatures reached 1800°F, setting the asphalt streets on fire. The hot winds swept across bomb shelters, suffocating and incinerating the occupants. Fifty thousand people died. The Hamburg police chief described it: "A fire typhoon such as was never before witnessed, against which every human resistance was quite useless."

---

of air from the blast, crushes structures and living things. The dynamic pressure, caused by the rushing shock front, is like a strong wind moving away from the explosion that pushes against every surface it encounters. The overpressure for a one-megaton surface burst is 10,000 pounds per square inch (psi) at about 700 feet from the hypocenter, 1000 psi at about 1500 feet, 100 psi at 2300 feet, 20 psi at 7000 feet, 10 psi at 10,000 feet, and 5 psi at about 3 miles. Frame houses collapse at 5 psi; reinforced concrete structures at 100 psi.

Where overpressures are 5 psi, the dynamic pressure is about 0.5 psi, which generates enough force to hurl a standing human being against a wall with several times the force of gravity. The same dynamic pressure causes debris, including glass shards, stones, metallic objects, and anything else shattered by the overpressure to fly with velocities above 100 miles per hour.

A one-megaton nuclear weapon exploding on ordinary soil at ground level will open a crater approximately 1300 feet across and 230 feet deep. The material ejected from the crater will cover the ground to a distance of 1800 to 2600 feet from the point of explosion.

At the same time, the explosion will result in ground shock that will generate earthquake-like tremors that can damage or collapse buildings several miles away. A 20-megaton weapon exploding on the ground will produce seismic energy equal to the energy of the 1906 San Francisco earthquake.

The atoms inside the early fireball are stripped of their electrons, and the gamma rays emitted by it knock off additional electrons from the surrounding atmosphere. All of these electrons fly out faster than the heavier, positively charged nuclei. Those electrons that move toward the earth are absorbed, thus creating an asymmetric charge distribution that gives rise to a giant electromagnetic pulse or EMP. This pulse of high voltage induces destructive current surges in electrical and electronic systems and so can burn out circuits over an area that extends for tens and even hundreds of miles beyond the point of detonation, incapacitating telephone and power lines, radios and televisions, electrical appliances, computers, and other components of the communications network and power grid. A large nuclear weapon exploded high above the earth could blanket an area one-half the size of the United States with destructive EMP.

Wood-frame house 3500 feet from ground zero, Operation Doorstep atomic test. Yucca Flats, Nevada, March 17, 1953. (U.S. Federal Emergency Management Agency.)

Operation Doorstep: a 15-kiloton atomic bomb explodes. (U.S. Federal Emergency Management Agency.)

Wood-frame house after Operation Doorstep (5-psi peak overpressure). (U.S. Federal Emergency Management Agency.)

curie originally referred to the radioactivity in 1 gram of radium, but now defined as the amount of radioactive material that undergoes 2.2×10^{12} radioactive disintegrations per minute.

hypocenter (or **ground zero**) the point on land directly below the center of an airburst nuclear explosion from which the blast, thermal, and radiation energies spread outward in all directions. The epicenter is the actual center of the explosion; for a surface burst the epicenter and hypocenter are essentially the same.

joule basic unit of energy. A flashlight battery has an energy value of about 10,000 joules.

rad (**r**adiation-**a**bsorbed-**d**ose) unit of radiation, defined as the absorption of 10^{-5} joule per gram of absorbing material.

rem (**r**oentgen-**e**quivalent-**m**an) amount of radiation required to produce the same biological effect as 1 roentgen (or rad) of moderate-energy gamma rays. For gamma rays, rems and rads are interchangeable.

Hiroshima Gas Company, a three-story reinforced concrete frame building, 720 feet from the hypocenter. The walls were 13 inches thick. (Hiroshima–Nagasaki Publishing Committee, U.S. Army returned materials.)

The total radioactivity generated by a one-megaton surface explosion is 10^{11} curies one hour after the detonation. It decays as shown in the figure on the next page. There is one-tenth of the original activity seven hours after detonation, one-hundredth 49 hours after, and one-thousandth approximately two weeks later. In contrast, the radioactivity released by the core meltdown of a one-gigawatt (1000 megawatts) nuclear reactor is initially about 10^8 curies, or one-thousandth that of a one-megaton weapon, but the radioactive materials released are significantly longer-lived, so that one-tenth of the original activity is found after approximately 1 year and one-hundredth after about 25 years.

If we look at the total land areas that would have to remain uninhabited because of the radioactive fallout following a one-megaton explosion, we see that an area of 13,000 square miles would be unusable for at least a week, if 10 rems per year were the maximum permitted accumulated dose. Of this 13,000 square miles, 120 square miles would be uninhabitable for at least a year at this same dose level. The U.S. Reactor Safety Study specifies 10 rems per year as the dose requiring evacuation because of the

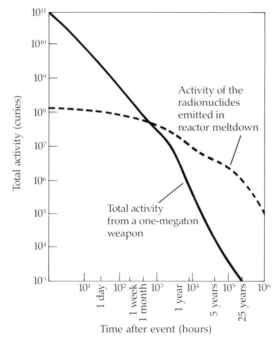

Radioactivity generated by a one-megaton nuclear weapon compared to the radioactivity released by a core meltdown of a one-gigawatt (1000 megawatts) nuclear reactor.

increased risk of cancer and leukemia above that dose. Chapter 18 treats the consequences of radioactive fallout in more detail, and depletion of the ozone layer is discussed in Chapter 19.

Some of the concepts described in this chapter may be difficult to comprehend, as they are beyond the experience and even the imagination of most people. The human dimensions of nuclear explosions for those at Hiroshima and Nagasaki and for those that would be victims of a present-day nuclear war will be examined in the following chapters.

SECTION II

Nuclear War, 1945

I am tired and sick of war. Its glory is all moonshine. It is only those who have neither fired a shot nor heard the shrieks and groans of the wounded, who cry aloud for blood, more vengeance, more desolation. War is hell.

General William Tecumseh Sherman
Michigan Military Academy
June 19, 1879

5

Nagasaki, August 9, 1945:
A Personal Account

Michito Ichimaru, M.D.

In August 1945, I was a freshman at Nagasaki Medical College.
The ninth of August was a clear, hot, beautiful summer day. I left
my lodging house, which was one and one-half miles from the
hypocenter, at eight in the morning, as usual, to catch a tram car.
When I got to the train stop, I found that it had been derailed in
an accident. I decided to return home. I was lucky. I never made it
to school that day.

At 11 AM, I was sitting in my room with a fellow student when
I heard the sound of a B-29 passing overhead. A few minutes
later, the air flashed a brilliant yellow and there was a huge blast
of wind.

We were terrified and ran downstairs to the toilet to hide. Later,
when I came to my senses, I noticed a hole had been blown in the
roof, all the glass had been shattered, and that the glass had cut
my shoulder and I was bleeding. When I went outside, the sky
had turned from blue to black and the black rain started to fall.
The stone walls between the houses were reduced to rubble.

After a short time, I tried to go to my medical school in
Urakami, which was 1500 feet from the hypocenter. The air dose
of radiation was more than 7000 rads at this distance. I could not
complete my journey because there were fires everywhere. I met
many people coming back from Urakami. Their clothes were in

43

rags, and shreds of skin hung from their bodies. They looked like ghosts with vacant stares. I cannot get rid of the sounds of crying women in the destroyed fields.

The next day I was able to enter Urakami on foot, and all that I knew had disappeared. Only the concrete and iron skeletons of the buildings remained. There were dead bodies everywhere. On each street corner we had tubs of water used for putting out fires after the air raids. In one of these small tubs, scarcely large enough for one person, was the body of a desperate man who had sought cool water. There was foam coming from his mouth, but he was not alive.

As I got nearer to school, there were black charred bodies, with the white edges of bones showing in the arms and legs. A dead horse with a bloated belly lay by the side of the road. Only the skeleton of the medical hospital remained standing. Because the school building was wood, it was completely destroyed. My class-mates were in that building, attending their physiology lecture. When I arrived, some were still alive. They were unable to move their bodies. The strongest were so weak that they were slumped over on the ground. I talked with them and they thought they would be OK, but all of them would die within weeks. I cannot forget the way their eyes looked at me and their voices spoke to me, forever.

I went up to the small hill behind the medical school, where all of the leaves of the trees were lost. The green mountain had changed to a bald mountain. There were many medical students, doctors, nurses, and some patients who escaped from the school and hospital. They were very weak and wanted water badly, cry-ing out, "Give me water, please." Their clothes were in rags, bloody and dirty. Their condition was very bad. I carried down several friends of mine on my back from this hill. I brought them to their houses, using a cart hitched to my bicycle. All of them died in the next few days. Some friends died with high fever, talking deliriously. Some friends complained of general malaise

---

rad   unit of radiation; an acute exposure of 450 rads will be lethal for approximately 50 percent of a healthy adult population, more than 600 rads will be lethal for virtually 100 percent.

Remains of a horse and cart near the hypocenter at Nagasaki. August 10, 1945. (Yosuke Yamahata, Hiroshima–Nagasaki Publishing Committee.)

Nagasaki Medical College was located on this hill, 1900 feet from the hypocenter. The wooden structures were demolished instantly and burned out. In the ruins of the five auditoriums the bodies of professors and students were found where they had been at the moment of the blast, on platforms and at desks. The total death toll, including the hospital, was 535 students, 42 of the teaching staff, 315 employees and nurses—892 in all. November 1945. (Hiroshima–Nagasaki Publishing Committee, U.S. Army returned materials.)

and bloody diarrhea, caused by necrosis of the bowel mucous membrane by severe radiation.

One of my jobs was to contact the families of the survivors. In all the public schools I visited, there were many many survivors brought there by the healthy people. It is impossible to describe the horrors I saw. I heard many voices in pain, crying out, and there was a terrible stench. I remember it as an inferno. All these people also died within several weeks.

One of my friends who was living in the same lodging house cycled back from medical school by himself that day. He was a strong man who did judo. That night he gradually became weak, but he went back to his home in the country by himself the next day. I heard he died a few weeks later. I lost many friends. So many people died that disposing of the bodies was difficult. We burned the bodies of my friends in a pile of wood which we gathered, in a small open place. I clearly remember the movement of the bowels in the fire.

On August 15, 1945, I left Nagasaki by train to return to my home in the country. There were many survivors in the same car. Even now, I think of the grief of the parents of my friends who died. I cannot capture the magnitude of the misery and horror I saw. Never again should these terrible nuclear weapons be used, no matter what happens. Only when mankind renounces the use of these nuclear weapons will the souls of my friends rest in peace.

6

Psychological Effects of the Atomic Bombings

Robert Jay Lifton, M.D.

The Experience Recalled

Since it was wartime, the people of Hiroshima expected conventional bombing, although only an occasional bomb actually had been dropped on the city. Air-raid warnings sounded regularly as planes passed overhead. At 8:15 AM on August 6, 1945, the moment the bomb fell, most people were in a particularly relaxed state. After a brief air-raid warning, the all-clear had just sounded.

Many people could not clearly recall their initial perceptions: many simply remember what they thought to be a flash—or else a sudden sensation of heat—followed by an indeterminate period of unconsciousness; others only recall being thrown across a room, or knocked down, then finding themselves pinned under the debris of buildings.

Adapted from "Psychological Effects of the Atomic Bomb in Hiroshima: The Theme of Death." Reprinted by permission of *Daedalus*, Journal of the American Academy of Arts and Sciences, 92 (3), Summer 1963, Boston, Mass. The photographs have been added.

Wrist watch stopped at 8:15 AM, the moment of the bombing. August 6, 1945.
(Yuichiro Sasaki, Hiroshima–Nagasaki Publishing Committee.)

A young shopkeeper's assistant, who was 13 years old at
the time the bomb fell, and 1400 meters (4500 feet) from the
hypocenter:

I was a little ill . . . so I stayed at home that day . . . . There had
been an air-raid warning and then an all-clear. I felt relieved
and lay down on the bed with my younger brother . . . . Then
it happened. It came very suddenly . . . it felt something like
an electric short—a bluish sparkling light . . . . There was a
noise, and I felt great heat—even inside of the house. When I

came to, I was underneath the destroyed house .... I didn't know anything about the atomic bomb so I thought that some bomb had fallen directly upon me . . . and then when I felt that our house had been directly hit I became furious .... There were roof tiles and walls—everything black—entirely covering me. So I screamed for help .... And from all around I heard moans and screaming, and then I felt a kind of danger to myself .... I thought that I too was going to die in that way. I felt this way at that moment because I was absolutely unable to do anything at all by my own power .... I didn't know where I was or what I was under .... I couldn't hear the voices of my family. I didn't know how I could be rescued. I felt I was going to suffocate and then die, without knowing exactly what had happened to me. This was the kind of expectation I had ....

The most striking psychological feature of this immediate experience was the sense of a sudden and absolute shift from normal existence to an overwhelming encounter with death, an emotional theme that remains with the victim indefinitely.

This early impact enveloped the city in an aura of weirdness and unreality. Only those at some distance from the bomb's hypocenter could clearly distinguish the sequence of the great flash of light in the sky accompanied by the lacerating heat of the fireball, followed by the sound and force of the blast, and then by the multicolored "mushroom cloud" rising above the city. Two thousand meters (6500 feet) is generally considered a crucial radius for high mortality (from heat, blast, and radiation) for a Hiroshima-size blast, for susceptibility to delayed radiation effects, and for near-total destruction of buildings and other structures. But many were killed outside of this radius, and indeed the number of deaths from the bomb—variously estimated from 63,000 to 240,000 or more—is still unknown. Falling in the center of a flat city made up largely of wooden residential and commercial structures, the bomb is reported to have destroyed or badly damaged (through blast and fire) more than two-thirds of all buildings within 5000 meters (3 miles), an area roughly encompassing the city limits, so that all of Hiroshima became immediately involved in the atomic disaster.

A middle-aged teacher who was on the outskirts of the city, about 5000 meters from the hypocenter, describes his awe at the destruction he witnessed:

Hiroshima, near a bridge 2.2 kilometers from the hypocenter, about 11:00 AM, August 6, 1945. Black smoke and raging flames shoot up from the heart of the city. Escaping the raging flames the people stand about, their seared skin hanging in strips, unable to go farther. They sit and lie by the bridge, filling every approach. The photographer, Yoshito Matsushige, said, "As I came near and raised my camera, my tears blurred the finder so that I could hardly see." (Hiroshima–Nagasaki Publishing Committee.)

I climbed Hijiyama Mountain and looked down. I saw that Hiroshima had disappeared . . . . I was shocked by the sight . . . . What I felt then and still feel now I just can't explain with words. Of course I saw many dreadful scenes after that—but that experience, looking down and finding nothing left of Hiroshima—was so shocking that I simply can't express what I felt. I could see Koi [a suburb at the opposite end of the city] and a few buildings standing . . . . But Hiroshima didn't exist—that was mainly what I saw—Hiroshima just didn't exist.

And a young university professor, 2500 meters from the hypo-center at the time, sums up these feelings of weird, awesome unreality in a frequently expressed image of hell:

> Everything I saw made a deep impression—a park nearby covered with dead bodies waiting to be cremated ... very badly injured people evacuated in my direction .... Perhaps the most impressive thing I saw were girls, very young girls, not only with their clothes torn off but their skin peeled off as well .... My immediate thought was that this was like the hell I had always read about .... I had never seen anything which resembled it before, but I thought that should there be a hell, this was it.

But human beings are unable to remain open to emotional experience of this intensity for any length of time, and very quickly—sometimes within minutes—people simply ceased to feel.

For instance, a male social worker in his twenties was at his home just outside the city; he rushed back into the city soon after the bomb fell. As one of the few able-bodied men left, he was put in charge of the work of disposing of corpses, which he found he could accomplish with little difficulty:

> After a while they became just like objects or goods that we handled in a very businesslike way .... Of course I didn't regard them simply as pieces of wood—they were dead bodies—but if we had been sentimental we couldn't have done the work .... We had no emotions .... Because of the succession of experiences I had been through, I was temporarily without feeling .... At times I went about the work with great energy, realizing that no one but myself could do it.

He contrasted his own feelings with the terror experienced by an outsider just entering the disaster area:

> Everything at the time was part of an extraordinary situation .... For instance, I remember that on the ninth or tenth of August, it was an extremely dark night ... I saw blue phosphorescent flames rising from the dead bodies—and there were plenty of them. These were quite different from the orange flames coming from the burning buildings .... These blue phosphorescent flames are what we Japanese look upon as spirits rising from dead bodies—in former days we called

Every day, cremations were carried out in the open air. Some were buried by their families, others by friends, many had no one to care for them. Here, a lonely, bereft family attends to the cremation of one of its members. Nagasaki, mid-September 1945. (Matsumoto Eiichi, Hiroshima–Nagasaki Publishing Committee.)

them fireballs—and yet at that time I had no sense of fear, not a bit, but merely thought "those dead bodies are still burning." . . . But to people who had just come from the outside, those flames looked very strange . . . . One of those nights I met a soldier who had just returned to the city, and I walked along with him . . . . He noticed these unusual fireballs and asked me what they were. I told him that they were the flames coming from dead bodies. The soldier suddenly became extremely frightened, fell down on the ground, and was unable to move . . . . Yet I at the time had a state of mind in which I feared nothing. Though if I were to see those flames now I might be quite frightened . . . .

Relatively few people were involved in the disposal of dead bodies. But virtually all the people I interviewed nonetheless experienced a similar form of psychic closing-off in response to their overall exposure to death. Many told how horrified they were when they first encountered corpses in strange array or extremely disfigured faces, but then, after a period of time as they saw more and more of these, they felt nothing. Psychic closing-off would last sometimes for a few hours and sometimes for days or even months and merge into long-term feelings of depression and despair.

But even the deep and unconscious psychological defensive maneuvers involved in psychic closing-off were ultimately unable to afford full protection to the survivor from the painful sights and stimuli impinging upon him. It was, moreover, a defense not devoid of its own psychological cost. Thus the same social worker, in a later interview, questioned his own use of the word "businesslike" to describe his attitude toward dead bodies and emphasized the pity and sympathy he felt while handling bodies and the pains he took to console family members who came for these remains; he even recalled feeling frightened at night when passing the spot where he worked at cremation by day. He was in effect telling me that not only was his psychic closing-off imperfect, but that he was horrified—felt ashamed and guilty—at having behaved in a way that he now thought callous. For he had indulged in activities that were ordinarily, for him, strongly taboo, and had done so with an energy, perhaps even an enthusiasm, that must have mobilized within him primitive emotions of a frightening nature.

The middle-aged teacher who had expressed such awe at the disappearance of Hiroshima reveals the way in which feelings of shame and guilt toward the dead break through the defense of psychic closing-off and painfully assert themselves:

> I went to look for my family. Somehow I became a pitiless person, because if I had pity I would not have been able to walk through the city, to walk over those dead bodies. The most impressive thing was the expression in peoples' eyes— bodies badly injured which had turned black—their eyes looking for someone to come and help them. They looked at me and knew I was stronger than they . . . . I was looking for my family and looking carefully at everyone I met to see if he or she was a family member—but the eyes—the emptiness— the helpless expression—were something I will never forget . . . . I often had to go to the same place more than once. I would wish that the same family would not still be there . . . .

I saw disappointment in their eyes. They looked at me with great expectation, staring right through me. It was very hard to be stared at by those eyes . . . .

He felt accused by the eyes of the anonymous dead and dying, of wrongdoing and transgression (a sense of guilt) for not helping them, for letting them die, for "selfishly" remaining alive and strong.

There were also many episodes of more focused guilt toward specific family members whom one was unable to help, and for whose death one felt responsible. For instance, the shopkeeper's assistant mentioned above was finally rescued from the debris of his destroyed house by his mother, but she was too weakened by her own injuries to be able to walk very far with him. Soon they were surrounded by fire, and he (a boy of 13) did not feel he had the strength to sustain her weight and became convinced that they would both die unless he took some other action. So he put her down and ran for help, but the neighbor he summoned could not get through to the woman because of the flames, and the boy learned shortly afterward that his mother died in precisely the place he had left her. His lasting sense of guilt was reflected in his frequent experience, from that time onward, of hearing his mother's voice ringing in his ears, calling for help.

A middle-aged businessman also related a guilt-stimulating sequence. His work had taken him briefly to the south of Japan, and he had returned to Hiroshima during the early morning hours of August 6. Having been up all night, he was not too responsive when his 12-year-old son came into his room to ask his father to remove a nail from his shoe so that he could put them on and go to school. The father, wishing to get the job quickly over with, placed a piece of leather above the tip of the nail and promised he would take the whole nail out when the boy returned in the afternoon. As in the case of many youngsters who were sent to factories to do "voluntary labor" as a substitute for their school work, the boy's body was never found—and the father, after a desperate, fruitless search for his son throughout the city, was left with the lingering self-accusation that the nail he had failed to remove might have impeded the boy's escape from the fire.

Most survivors focused upon one incident, one sight, or one particular *ultimate horror* with which they strongly identified

This aged woman seems to have lost the use of her legs, probably due to the shock. Her position suggests a loss of all sense of reality. Nagasaki, before noon, August 10, 1945, about 1.3 kilometers south of the hypocenter. (Yosuke Yamahata, Hiroshima–Nagasaki Publishing Committee.)

themselves, and which left them with a profound sense of pity, guilt, and shame. Thus, the social worker describes an event that he feels affected him even more than his crematory activities:

On the evening of August 6, the city was so hot from the fire that I could not easily enter it, but I finally managed to do so by taking a path along the river. As I walked along the bank near the present Yokogawa Bridge, I saw the bodies of a mother and her child .... That is, I thought I saw dead bodies, but the child was still alive—still breathing, though with difficulty .... I filled the cover of my lunch box with water and gave it to the child but it was so weak it could not drink. I knew that people were frequently passing that spot . . . and I hoped that one of these people would take the child—as I had to go back to my own unit. Of course I helped many people all through that day . . . but the image of this child stayed on my mind and remains as a strong impression even now . . . . Later when I was again in that same area I hoped that I might be able to find the child . . . and I looked for it among all the dead children collected at a place nearby . . . . Even before the war I planned to go into social work, but this experience led me to go into my present work with children—as the memory of that mother and child by Yokogawa Bridge has never left me, especially since the child was still alive when I saw it.

These expressions of ultimate horror can be related to direct personal experience of loss (for instance, the businessman who had failed to remove the nail from his son's shoe remained preoccupied with pathetic children staring imploringly at him), as well as to enduring individual emotional themes. Most of them involved women and children, universal symbols of purity and vulnerability, particularly in Japanese culture. And, inevitably, the ultimate horror was directly related to death or dying.

Contamination and Disease

Survivors told me of three rumors that circulated widely in Hiroshima just after the bomb. The first was that for a period of 75 years Hiroshima would be uninhabitable: no one would be able to live there. This rumor was a direct expression of the *fear of deadly and protracted contamination from a mysterious poison believed to have been emitted by the frightening new weapon*. (As one survivor put it, "The ordinary people spoke of poison, the intellectuals spoke of radiation.")

Even more frequently expressed, and I believe with greater emotion, was a second rumor: trees and grass would never again grow in Hiroshima; from that day on, the city would be unable to sustain vegetation of any kind. This rumor seemed to suggest *an ultimate form of desolation even beyond that of human death:* nature was drying up altogether, the ultimate source of life was being extinguished—a form of symbolism particularly powerful in Japanese culture with its focus upon natural aesthetics and its view of nature as both enveloping and energizing all of human life.

The third rumor, less frequently mentioned to me but one that also had wide currency in various versions, was that all those who had been exposed to the bomb in Hiroshima would be dead within three years. This more naked death symbolism was directly related to the appearance of frightening symptoms of toxic radiation effects. For almost immediately after the bomb and during the following days and weeks, people began to experience, and notice in others, symptoms of a strange form of illness: nausea, vomiting, and loss of appetite; diarrhea with large amounts of blood in the stools; fever and weakness; purple spots on various parts of the body from bleeding into the skin; inflammation and ulceration of the mouth, throat, and gums; bleeding from the mouth, gums, nose, throat, rectum, and urinary tract; loss of hair from the scalp and other parts of the body; extremely low white blood cell counts when these were taken; and in many cases a progressive course toward death. These symptoms and fatalities aroused in the minds of the people of Hiroshima a special terror, *an image of a weapon that not only kills and destroys on a colossal scale but also leaves behind in the bodies of those exposed to it deadly influences that may emerge at any time and strike down their victims.* This image was made particularly vivid by the delayed appearance of these radiation effects, two to four weeks after the bomb fell, sometimes in people who had previously seemed to be in perfect health.

The shopkeeper's assistant, both of whose parents were killed by the bomb, describes his reactions to the death of two additional close family members from these toxic radiation effects:

My grandmother was taking care of my younger brother on the 14th of August when I left, and when I returned on the 15th she had many spots all over her body. Two or three days later she died . . . . My younger brother, who . . . was just a [five-month-old] baby, was without breast milk—so we fed him thin rice gruel . . . . But on the 10th of October he suddenly began to look very ill, though I had not then noticed any spots on his body . . . . Then on the next day he began to look a little better, and I thought he was going to survive. I was very pleased, as he was the only family member I had left, and I took him to a doctor—but on the way to the doctor he died. And at the time we found that there were two large spots on his bottom . . . . I heard it said that all these people would die within three years . . . so I thought, "sooner or later I too will die." . . . I felt weak and very lonely—with no hope at all . . . and since I had seen so many people's eyebrows falling out, their hair falling out, bleeding from their teeth—I found myself always nervously touching my hair like this [he demonstrated by rubbing his head] . . . . I never knew when some sign of the disease would show itself . . . . And living in the countryside then with my relatives, people who came to visit would tell us these things and then the villagers also talked about them—telling stories of this man or that man who visited us a few days ago, returned to Hiroshima, and died within a week . . . . I couldn't tell whether these stories were true or not, but I believed them then. And I also heard that when the *hibakusha* were evacuated to the village where I was, they died there one by one . . . . This loneliness, and the fear . . . the physical fear . . . has been with me always . . . . It is not something temporary, as I still have it now . . . .

Here we find a link between this early sense of ubiquitous death from radiation effects and later anxieties about death and illness. In a similar tone, a middle-aged writer describes his daughter's sudden illness and death:

My daughter was working with her classmates at a place 1000 meters from the hypocenter . . . . I was able to meet her the next day at a friend's house. She had no burns and only minor external wounds so I took her with me to my country house. She was quite all right for a while but on the 4th of September she suddenly became sick . . . . The symptoms of her disease were different from those of a normal disease . . . . She had spots all over her body . . . . Her hair began to fall out. She vomited small clumps of blood many times. Finally she began to bleed all over her mouth. And at times her fever was very high. I felt this was a very strange and horrible disease . . . . We didn't know what it was. I thought it was a kind of epidemic—something like cholera. So I told the rest

---

hibakusha   the Japanese word for those who experienced the atomic bomb, literally "explosion-affected person or persons."

A boy carrying his younger brother on his back. His face is covered with dried blood. Their parents probably have been killed. Nagasaki, about 7 AM, August 10, 1945, 2.3 kilometers from the hypocenter. (Yosuke Yamahata, Hiroshima–Nagasaki Publishing Committee.)

of my family not to touch her and to disinfect all utensils and everything she used . . . . We were all afraid of it and even the doctor didn't know what it was . . . . After ten days of agony and torture she died on September 14 . . . . I thought it was very cruel that my daughter, who had nothing to do with the war, had to be killed in this way . . . .

Survivors were thus affected not only by the fact of people dying around them but by the way in which they died: a gruesome form of rapid bodily deterioration that seemed unrelated to more usual and "decent" forms of death.

We have seen how these initial physical fears could readily turn into lifetime bodily concerns. And during the years that followed, these fears and concerns became greatly magnified by another development: the growing awareness among the people of Hiroshima that medical studies were demonstrating an abnormally high rate of leukemia among survivors of the atomic bomb. The increased incidence was first noted in 1948 and reached a peak between 1950 and 1952; it has been greatest in those exposed closest to the hypocenter so that for those within 1000 meters the increase of leukemia has been between 10 and 50 times the normal. Since 1952 the rate has diminished, but it is still higher than in unexposed populations, and fears remain strong. While symptoms of leukemia are not exactly the same as those of acute radiation effects, the two conditions share enough in common—the dreaded "purple spots" and other forms of hemorrhage, laboratory findings of abnormalities of the blood, progressive weakness and fever, and (inevitably in leukemia, and often enough in acute irradiation) ultimate death—that these tend to merge, psychologically speaking, into a diffuse fear of bodily annihilation and death.

Moreover, Hiroshima survivors are aware of the general concern and controversy about genetic effects of the atomic bomb, and most express fear about possible harmful effects upon subsequent generations—a very serious emotional concern anywhere, but particularly so in an East Asian culture that stresses family lineage and the continuity of generations as man's central purpose in life and (at least symbolically) his means of achieving immortality. The Hiroshima people know that radiation *can* produce congenital abnormalities (as has been widely demonstrated in laboratory animals), and abnormalities have frequently been reported among the offspring of survivors.

Fears about general health and genetic effects have affected marriage arrangements (which are usually made in Japan by families with the help of a go-between). Survivors encounter discrimination, particularly when involved in arrangements with families outside of Hiroshima.

A company employee in his thirties, who was 2000 meters from the bomb's hypocenter when it fell, described to me virtually all these bodily and genetic concerns in a voice that betrayed considerable anxiety:

Even when I have an illness which is not at all serious—as for instance when I had very mild liver trouble—I have fears about its cause. Of course if it is just an ordinary condition there is nothing to worry about, but if it has a direct connection to radioactivity, then I might not be able to expect to recover. At such times I feel myself very delicate . . . . But in general, there is a great concern that people who were exposed to the bomb might become ill five or ten years later or at any time in the future . . . . Also when my children were born, I found myself worrying about things that ordinary people don't worry about, such as the possibility that they might have inherited some terrible disease from me . . . . I heard that the likelihood of our giving birth to deformed children is greater than in the case of ordinary people . . . and at that time my white blood cell count was rather low . . . . I felt fatigue in the summertime and had a blood count done three or four times . . . . I was afraid it could be related to the bomb, and was greatly worried . . . . Then after the child was born, even though he wasn't a deformed child, I still worried that something might happen to him afterward . . . . With the second child too I was not entirely free of such worries . . . . I am still not sure what might happen, and I worry that the effects of radioactivity might be lingering in some way . . . .

Here we see a young man carrying on effectively in his life, essentially healthy, with normal children, and yet continually plagued by underlying anxieties about his general health and about the birth and health of his children. Each hurdle is passed, but there is little relief; like many survivors, he experiences an inner sense of being doomed.

And a young clerk, also exposed about 2000 meters from the hypocenter, but having the additional disadvantage of a keloid scar resulting from facial burns, expresses similar emotions in still stronger fashion:

Frankly speaking, even now I have fear . . . . Even today people die in the hospitals from A-bomb disease, and when I hear about this I worry that I too might sooner or later have the same thing happen to me . . . . I have a special feeling that I am different from ordinary people . . . that I have the mark of wounds—as if I were a cripple . . . . I imagine a person who has an arm or a leg missing might feel the same way . . . . It is not a matter of lacking something externally, but rather something like a handicap—something mental that does not show—the feeling that I am mentally different from ordinary people . . . so when I hear about people who die from A-bomb disease or who have operations because of this illness, then I feel that I am the same kind of person as they . . . .

The survivor's identification with the dead and the maimed initiates a vicious circle on the psychosomatic plane of existence: he is likely to associate the mildest everyday injury or sickness with possible radiation effects, and anything he relates to radiation effects becomes associated with death. The process is accentuated by the strong Japanese cultural focus upon bodily symptoms as expressions of anxiety and conflict. Thus the all-encompassing term "atomic bomb sickness" or "atomic bomb disease" (*genbakushō*) has evolved, referring on the one hand to such fatal conditions as the early acute radiation effects and later cases of leukemia and on the other hand to the vague, borderline area of fatigue, general weakness, sensitivity to hot weather, suspected anemia, susceptibility to colds or stomach trouble, and general nervousness—all of which are frequent complaints of survivors, and which many associate with radiation effects.

Doctors are caught in a conflict between humanitarian provision for medical need and the danger of encouraging the development in survivors of hypochondria, general weakness, and dependency—or what is sometimes called "atomic bomb neurosis." Since *genbakushō* is at this historical juncture as much a spiritual as a physical condition (as the young clerk made so clear)—and one that touches at every point upon the problem of death—it is difficult for any law or medical program to provide a cure.

Unwanted Identity

It is clear by now that exposure to the atomic bomb changed the survivor's status as a human being, in his own eyes as well as those of others. Both through his immediate experience and its consequences over the years, he became a member of a new group; he assumed the identity of the *hibakusha*, of one who has undergone the atomic bomb. When I asked survivors to associate freely to the word *hibakusha* and to explain their feelings about it, they invariably conveyed to me the sense of having been compelled to take on this special category of existence, by which they felt permanently bound, however they might wish to free themselves from it. The shopkeeper's assistant expresses this in simple terms characteristic for many:

Well . . . because I am a *hibakusha* . . . how shall I say it—I wish others would not look at me with special eyes . . . perhaps *hibakusha* are mentally—or both physically and mentally—different from others . . . but I myself do not want to be treated in any special way because I am a *hibakusha* . . . .

To be a *hibakusha* thus separates one from the rest of humankind. It means, as expressed by a young female clerical worker left with a keloid from her atomic-bomb exposure at 1600 meters, a sense of having been forsaken.

I don't like people to use that word [*hibakusha*] . . . . Of course there are some who, through being considered *hibakusha*, want to receive special coddling [*amaeru*] . . . . But I like to stand up as an individual. When I was younger they used to call us "atomic bomb maidens." . . . More recently they call us *hibakusha* . . . . I don't like this special view of us . . . . Usually when people refer to young girls, they will say girls or daughters, or some person's daughter . . . but to refer to us as atomic bomb maidens is a way of discrimination . . . . It is a way of abandoning us . . . .

What she is saying, and what many said to me in different ways, is that the experience, with all its consequences, is so profound that it can virtually become the person. Others then see the person *only* as a *hibakusha* bearing the taint of death and therefore, in the deepest sense, turn away. And even the special attentions, the various forms of emotional succor that the survivor may be tempted to seek, cannot be satisfying because such succor is ultimately perceived as inauthentic.

A European priest, one of the relatively few non-Japanese *hibakusha*, expresses these sentiments gently but sardonically:

I always say—if everyone looks at me because I received the Nobel Prize, that's OK, but if my only virtue is that I was 1000 meters from the atomic bomb center and I'm still alive—I don't want to be famous for that.

However well or poorly a survivor is functioning in his life, the word *hibakusha* evokes an image of the dead and the dying. The young clerk, for instance, when he hears the word, thinks either

Hibakusha couple—injured and homeless. Note the numerous flies on their shirts. Hiroshima, 1945. (Jay Eyerman, Life Magazine. © *Time, Inc.)*

of the experience itself ("Although I wasn't myself too badly injured I saw many people who were . . . and I think . . . of the look on their faces . . . camps full of these people, their breasts burned and red") or, as we have already heard him describe, of the aftereffects ("When I hear about people who die from *genbakushō* or who have operations because of this illness, then I feel that I am the same kind of person as they").

We are again confronted with the survivor's intimate identification with the dead; we find, in fact, that it tends to pervade the entire *hibakusha* identity. *For survivors seem not only to have experienced the atomic disaster, but to have imbibed it and incorporated it into their beings, including all of its elements of horror, evil, and particularly of death.* They feel compelled virtually to merge with those who died, not only with close family members but with a more anonymous group of "the dead."

The *hibakusha* identity, then, in a significant symbolic sense, becomes an identity of the dead. Created partly by the particularly intense Japanese capacity for identification, and partly by the special quality of guilt over surviving, it takes shape through the following inner sequence: I almost died; I should have died; I did die, or at least am not really alive; if I am alive, it is impure of me to be so; and anything I do that affirms life is also impure and an insult to the dead, who alone are pure.

Finally, this imposed identity of the atomic bomb survivor is greatly affected by his historical perceptions (whether clear or fragmentary) of the original experience, including its bearing upon the present world situation. The dominant emotion is the survivors' sense of having been guinea pigs, not only because research groups (particularly American research groups) interested in determining the effects of delayed radiation have studied them, but more fundamentally because they were the victims of the first "experiment" (a word many of them use in referring to the event) with nuclear weapons. They are affected by a realization, articulated in various ways, that they have experienced something ultimate in man-made disasters, and at the same time by the feeling that the world's continuing development and testing of the offending weapons deprive their experience of meaning.

Beyond Hiroshima

Does Hiroshima follow the standard patterns delineated for other disasters, or is it—in an experiential sense—a new order of event? The usual emotional patterns of disaster are very much present. We can break down the experience into the usual sequence of anticipation, impact, and aftermath. We can recognize such standard individual psychological features as various forms of denial, the "illusion of centrality" (the feeling of each person that he was at the very center of the disaster's path), the apathy of the "disaster syndrome" resulting from the sudden loss of the sense of safety and even omnipotence with which we usually conduct our lives, and the conflict between self-preservation and wider human responsibility, which culminates in feelings of guilt and shame. Even some of the later social and psychological conflicts in the affected population are familiar. Yet we have also seen convincing evidence that the Hiroshima experience, no less in the psychological than in the physical sphere, transcends the ordinary disaster. Hiroshima was and is both a direct continuation of the long and checkered history of human struggle and a plunge into a new and tragic dimension.

When we consider the turbulent onset of an encounter with death at the moment the bomb fell, its shocking reappearance in association with delayed radiation effects, and its prolonged expression in the group identity of the doomed and nearly dead, we are struck by the fact that it is an interminable encounter. There is, psychologically speaking, no end point, no resolution. This continuous and unresolvable encounter with death, then, is a unique feature of the atomic bomb disaster.

A psychological feature of particular importance in the Hiroshima disaster is what I have called *psychic numbing*. Resembling the psychological defense of denial and the behavioral state of apathy, psychic numbing is nonetheless a distinctive pattern of response to overwhelmingly threatening stimuli. Within a matter of moments, as we have seen in the examples cited, a person may not only cease to react to these threatening stimuli, but in so doing may, equally suddenly, violate the most profound values and taboos of his cultural and personal life. Though the response may be highly adaptive—and may very often be a means of emotional self-preservation—it can vary in its proportions to the extent of almost resembling at times a psychotic mechanism. The psychic numbing created by the Hiroshima disaster is not limited to the victims themselves but extends to those who attempt to study the event.

Our inability to imagine death, the elaborate circle of denial, and the profound inner need of human beings to make believe they will never die are universal psychological barriers to thought about death. The enormity of the scale of killing in a nuclear disaster and the impersonal nature of the technology are still further impediments to comprehension.

No wonder, then, that the world resists full knowledge of the Hiroshima and Nagasaki experiences and expends relatively little energy in comprehending their full significance. And beyond Hiroshima, these same impediments tragically block and distort our perceptions of the general consequences of nuclear weapons.

This volume, detailing the human tragedy at Hiroshima and Nagasaki and warning of the devastation that would result from a modern nuclear war, has been designed to help the reader overcome this resistance. Only those who permit themselves to confront the reality of nuclear weapons effects, past or future, can begin to grasp the danger we now face. Human survival may well depend on this ability.

7

Acute Medical Effects
at Hiroshima and Nagasaki

Takeshi Ohkita, M.D.

Modern science and technology have brought us many hopes
and dreams. At the same time, they have caused much anxiety.
From their sufferings since the atomic bombings in August 1945,
the Japanese people recognize that the day has come when nuclear energy could be the weapon of ultimate destruction. Human intelligence has discovered something that is a cause for
grave concern.

The clear marks left on the somatic cells of human bodies have
not disappeared after more than 35 years. They are still causing
various disturbances, which you will read about in the following pages.

I believe that we, as doctors and scientists, once again must
realize the importance of our responsibilities. Together we must
gather our wisdom and intelligence to abolish nuclear weapons
from the face of the earth.

The acute effects of the Hiroshima and Nagasaki atomic bombs
are summarized here, based on documentary records. Acute injuries caused by the atomic bombs are classified as thermal, mechan-

Adapted from "Review of Thirty Years' Study of Hiroshima and Nagasaki
Atomic Bomb Survivors," *Journal of Radiation Research*, Supplement, 49–66, 1975.
The photographs have been added.

Nagasaki, August 9, 1945, three minutes after the bombing. The smoke column has reached 20,000 feet. (Hiroshima–Nagasaki Publishing Committee, U.S. Army returned materials.)

ical, and radiation injuries. Combinations of these were most common. Many people died from the immediate effects of blast and burns, but individuals often succumbed to trauma or burns before the radiation syndrome developed. Many more would have died from irradiation, had they been saved from the effects of trauma or burns. Nearly all who died within ten weeks had signs suggestive of radiation injuries. Remarkable variation in sensitivity of body tissues to ionizing radiation was apparent. Radiation-induced bone marrow depletion was the most critical damage leading to death. In these instances, lowered numbers of white blood cells and platelets and subsequent infections and hemorrhagic tendencies were the main causes of death.

. . .

Thermal Injuries

The intensity of the heat generated by the nuclear explosions in Japan is estimated to have been 3000–4000°C (5400–7200°F) at ground level near the hypocenters. Its duration was exceedingly short—0.5 to 1 second. The heat markedly dissipated with increasing distances from the hypocenters, but there was evidence that it was more than 573°C (1030°F) at distances of 1000 to 1100 meters (3200–3610 feet) and 1600 meters (5250 feet) from the hypocenters in Hiroshima and Nagasaki, respectively.

Thermal radiation causes burns directly, or indirectly from fires started by the flash. Direct burns are often called "flash burns," since they are produced by the flash of thermal radiation from the fireball. Everyone exposed unshielded within 4 kilometers (2.5 miles) of the hypocenters (ground zero) probably received burns of some degree. Those beneath the burst were burned to death. In addition, persons in buildings close to the hypocenters might have been burned by hot gases and dust entering the structures, even though they were adequately shielded from direct thermal radiation. Severe third-degree burns with charring and skin death were commonly observed among people who were in the open within 1 kilometer (.6 mile) of the hypocenters.

Flash burns have also been termed "profile burns" since the lesions occur on the unshielded parts of the body exposed in a direct line with the origin of the thermal radiation. They were usually restricted to one side of the body and were sharply out-

Charred boy 700 meters from the hypocenter, Nagasaki. (Yosuke Yamahata, Hiroshima–Nagasaki Publishing Committee.)

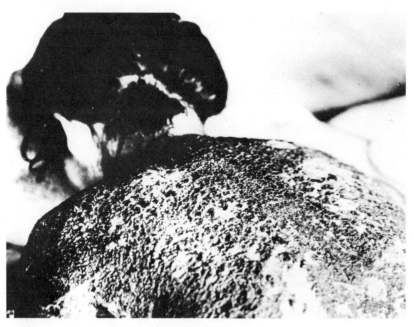

Woman with severely burned back, Hiroshima, August 7, 1945. (Masayoshi Onuka, Hiroshima–Nagasaki Publishing Committee.)

lined. At the times of the bombings, hot weather prevailed. Most people wore short-sleeved shirts without coats. The effects of radiant heat were enhanced on the bare skin since clothing was protective to a variable degree, depending on its quality and color and the intensity of the heat.

Indirect burns, referred to as "flame burns," are identical with skin burns caused by fire. They may involve any or all parts of the body and tend to penetrate much deeper than do "flash burns." There were no essential differences in the healing processes of these two types of burns.

The frequency of burn injuries was exceptionally high. Burns seem to have been the major cause of death on the days of the bombings, but their relative proportion among all deaths is unknown. Many who were injured by the blasts were unable to escape and died in the fires. Flash and flame burns were often combined—some people were burned when their clothes were ignited by the flash of heat. Burns occurred under clothing at least as far as 2.5 kilometers (1.5 miles) from the hypocenters. The burns of those who survived, however, were largely flash burns. The incidence of flame burns appears to have been very small, constituting no more than 5 percent of the total burns.

All those in the open air without appreciable protection received severe burns within 1.5 kilometers of the hypocenters; moderate but fatal burns within 2.5 kilometers; and mild burns at distances of 3 to 4 kilometers from the hypocenters. As shown in Table 1, the incidence of burns in Hiroshima was nearly 100 percent among unshielded survivors at distances up to 2.5 kilometers, beyond which it fell rapidly. Burns were most frequent in persons outdoors and unshielded, considerably less frequent in

---

Parents, half crazy with grief, searched for their children. Husbands looked for their wives, and children for their parents. One poor woman, insane with anxiety, walked aimlessly here and there through the hospital calling her child's name.

M. Hachiya, M.D.
Hiroshima Diary: The Journal of a Japanese Physician,
August 6–September 30, 1945

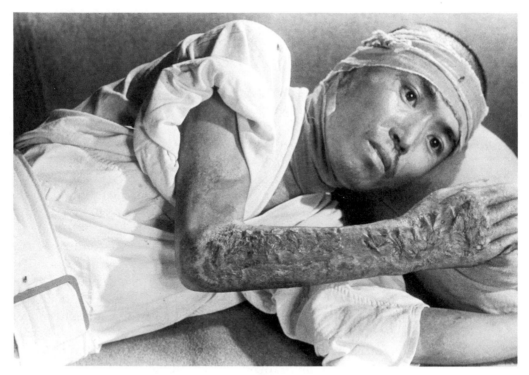

Flash burns at Hiroshima, 2 kilometers from the hypocenter. This man was wearing short sleeves when the bomb exploded. He had a skin graft, taken from his hip, on his right hand. October 2, 1945. (Shunkichi Kikuchi, Hiroshima–Nagasaki Publishing Committee.)

those outdoors and shielded, and least among those who were indoors. There was little difference in the incidence of burns among those in concrete buildings versus those in Japanese-style houses.

The thermal energy was estimated to have been somewhat higher in Nagasaki, but the prevalence of burns was much lower. The overall incidence of flash burns in Nagasaki by distance from the hypocenter was similar to that of Hiroshima. A few second-degree burns with reddening of skin and blistering were recorded at distances from the hypocenters of 3.3 kilometers in Hiroshima and 3.1 kilometers in Nagasaki. No flash burns were reported beyond 4 kilometers in Hiroshima; whereas, in Nagasaki, about 3 percent of the persons exposed at 4 to 5 kilometers were reported to have received first-degree flash burns. The influence of shielding on the incidence of burns is clearly demon-

Table 1. Burns by distance and shielding in individuals living 60 days after the bombing, Hiroshima.

Distance from Hypocenter (km)	Shielding Conditions				Total Burns by Degree (%)		
	Outdoors Unshielded	Outdoors Shielded	Indoors		First	Second	Third
	Cases with Burns / Number Investigated (%)						
0–0.5		2/3 (66.6)	3/24 (12.5)		6.2	52.7	41.1
0.6–1.0	22/22 (100.0)	34/68 (50.0)	33/210 (15.7)				
1.1–1.5	172/172 (100.0)	50/144 (34.7)	105/631 (16.6)				
1.6–2.0	518/528 (98.1)	64/176 (36.3)	135/770 (17.5)				
2.1–2.5	439/443 (99.0)	69/150 (46.0)	50/563 (8.8)				
2.6–3.0	98/124 (79.0)	19/94 (20.2)	23/284 (8.0)				
3.1–3.5	33/85 (38.8)	2/58 (3.4)	6/230 (2.6)				
3.6–4.0	4/40 (10.0)	0/12	0/102	Total of all conditions			
Total	1286/1414 (90.9)	240/705 (34.0)	355/2814 (12.6)	1881/4933 (38.1)			

Source: T. Kajitani and S. Hatano in Collection of the Reports, 1953; and Report of the Joint Commission, 1951.

strated from the data in Table 1. People inside buildings were burned only when the rays could reach them through doors and windows.

The incidence of burns was low up to 1.5 kilometers from the hypocenters due to the high mortality in this group. Survivors within this distance were probably partially shielded against radiant heat. Such a selective factor could account for the lower incidence of burns among survivors in an area where they otherwise

Collapsed stairways trapped students at the Shiroyama Elementary School in Nagasaki. (Hiroshima–Nagasaki Publishing Committee, U.S. Army returned materials.)

could be expected to have been fatal. After the healing of severe burns, overgrowth of scar tissue was frequently observed, especially among survivors who were burned within 2.5 kilometers from the hypocenters [see Chapter 8]. Appreciable regression of protruding scars had occurred in most cases by 1952.

Mechanical (Blast) Injuries

The blast pressures generated by the Hiroshima and Nagasaki atomic bombs at ground zero are estimated to have been 4.5–6.7 and 6–8 tons per square meter, respectively. The blasts consisted of two phases: compression and suction. The duration of the compression phase is estimated to have been approximately one-half to one second. Mechanical injuries resulting from the blasts were direct and indirect, mostly the latter, and were chiefly caused by collapsing buildings and flying debris.

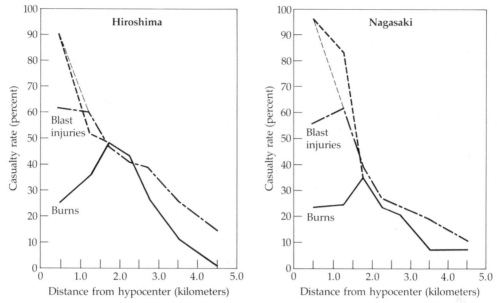

Incidence of blast injuries and burns by distance. The dotted lines show the probable casualty rates in the innermost zones, assuming that all those who died were injured by blast and radiant heat. (After Oughterson and Warren, 1956.)

So far, there have been no reliably established deaths attributable to direct blast effects. [Editors' note: That is, resulting from the direct blast pressure on human bodies alone, exclusive of collisions between bodies and various objects.] Indirectly, the blasts caused many instantaneous deaths. The incidence of indirect mechanical injuries among survivors is shown in the figure above. Blast injuries occurred mostly among people in concrete buildings, somewhat less among those in Japanese-style houses, less outdoors with shielding, and least frequently among those outdoors without shielding—exactly the reverse order from that of burns. This also suggests that buildings and walls offered more risk than protection, especially at close range.

Mechanical injuries of survivors were of all degrees, from minor scratches to severe lacerations and compound fractures. The most common injury was laceration by small glass fragments. Fractures were infrequent, but many who did not survive probably had severe fractures. With the extreme scarcity of medical care soon after the bombings and because of leukopenias (lowered white blood cell counts) due to ionizing radiation, minor lacerations and abrasions, which ordinarily would have promptly healed, often resulted in severe infections.

Rupture of eardrums was considered evidence of direct blast injury among survivors. At the Ninoshima Hospital in Hiroshima, soon after the atomic bomb, only 8 (2 percent) of 371 patients examined had ruptured eardrums, though 19 of the 371 were temporarily deaf. Seventy-six percent of the 371 had been within 2 kilometers of the hypocenter. Eight percent of 198 Nagasaki survivors exposed within 1 kilometer and examined in October 1945 had ruptured eardrums. Among survivors in both cities surveyed by questionnaire, less than 1 percent at any distance reported this condition, and none who were located farther than 3 kilometers reported it.

Other less-defined symptoms may have been blast-related, such as vertigo, tinnitus (buzzing or ringing in the ears), and headache—without evidence of trauma. About 15 percent of the survivors surveyed in each city complained of these symptoms. Most of them had been within 2.5 kilometers. Loss of consciousness was also reported, but most cases of transitory unconsciousness were more likely caused by violent displacement, such as being thrown to the ground, rather than by direct blast. Regarding later effects of mechanical injuries, no data for the precise number of disabled survivors are available.

Radiation Injuries

Though the Hiroshima and Nagasaki atomic bombs afforded the first opportunity to observe the effects of massive ionizing radiation exposure in humans, little is known of the severe radiation injuries that caused immediate deaths, because these cases were not autopsied. In addition, the high rates of deaths and injuries during the first few days after the atomic bombs precluded an accurate statistical evaluation of the effects of ionizing radiation. The symptoms among survivors with radiation injuries and alive three or more weeks after the bombs will be briefly presented. Bear in mind that definitive criteria for diagnosing radiation injuries and their severity are difficult to establish because such injuries (e.g., leukopenia and thrombocytopenia) may not have been immediately manifest in many cases, and some symptoms may have been attributed to or complicated by causes (e.g., burns, mechanical injuries, poor sanitary conditions) other than radiation.

78

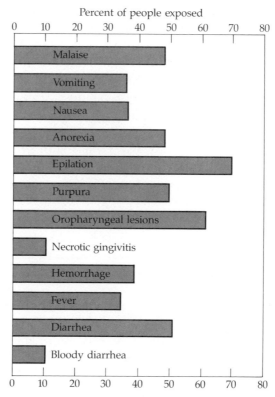

Percent of people exposed

Prevalence of symptoms and signs within 1000 meters of the hypocenter at Hiroshima. The air dose at 1000 meters was 447 rads. Diarrhea includes bloody diarrhea. (After the Report of the Joint Commission, 1951.)

Clinical Symptoms and Signs of Radiation Injuries

The prevalence of the main clinical symptoms and signs in those located within 1.0 kilometer of the hypocenter in Hiroshima is shown in the figure above. These survivors were undoubtedly exposed to large doses of ionizing radiation except for a few who were heavily shielded, as by bomb shelters. Epilation (the loss of hair), purpura (hemorrhaging into the skin or internal organs), and oropharyngeal ulceration (punched-out sores in the back of the throat) were the most frequent signs. The first two were usually in combination and are considered the most reliable diagnostic signs of radiation injury. Oropharyngeal ulceration was not as specific for radiation injury.

Chronologically, symptoms and signs can be grouped as fol-

79

lows: Phase I, the prodromal (the early manifestations of an illness) radiation syndrome, usually of one or more days' duration and consisting of prostration and gastrointestinal symptoms, including nausea, vomiting, and anorexia (loss of appetite); Phase II, a period of relative well-being whose duration is variable but inversely proportional to the exposure dose; Phase III, a feverish period of several weeks' duration, with epilation, oropharyngeal ulceration, infection, hemorrhaging, and diarrhea; and Phase IV, either death or prolonged convalescence with eventual recovery.

In many cases, the second phase of temporary well-being was not observed. In those severely injured, the feverish phase often began between the fifth and seventh days, but sometimes as early

---

hematological pertaining to the blood and blood-forming tissues.

hematopoietic pertaining to the formation of blood cells.

leukopenia lowered white blood cell count. Normal values are 5000–10,000 white blood cells per cubic millimeter of blood.

prodromal radiation syndrome Phase I of radiation illness—the early manifestations of radiation injury.

rad unit of radiation; an acute exposure of 450 rads will be lethal for approximately 50 percent of a healthy adult population, more than 600 rads will be lethal for virtually 100 percent.

Relative Biological Effectiveness (RBE) for a given type and energy of radiation, the dose of gamma rays necessary to produce the same biological effect as a unit dose of the radiation in question. The RBE of gamma rays, the reference, is 1; beta rays are also roughly 1; thermal neutrons are 5; fast neutrons 10; and alpha particles 20. Therefore, 1 rad of fast neutrons produces the same biological effect in humans (rem) as 10 rads of gamma rays.

rem (roentgen-equivalent-man) amount of radiation that, when absorbed into the body, produces the same biological effect as 1 roentgen (or rad) of gamma rays.

thrombocytopenia lowered blood platelet count. Normal values are 200,000–500,000 platelets per cubic millimeter of blood.

---

as the third—with severe diarrhea as its most prominent manifestation—and continued until death. In the less severely injured, epilation about two weeks after exposure—initiating the feverish phase—was soon followed by purpura and oropharyngeal lesions. Despite marked individual variation, the severity of manifestations depended on the radiation exposure. The severely exposed (as great as 450 rads or more) usually died within two weeks. Less severely but fatally exposed died, as a rule, before the end of the sixth or eighth week after exposure. Additional comments about the symptoms and signs of radiation injuries follow.

Nausea and Vomiting. Though there were many causes other than ionizing radiation for nausea and vomiting, these are well-established initial signs of radiation sickness. Distance correlated closely with vomiting on the first day; less so, later. In the more heavily exposed, these signs usually persisted, frequently lasting several days, and in some throughout the entire course of illness. The presence of burns had no influence on nausea and vomiting.

Epilation. Epilation is considered one of the specific signs of radiation injury. It began one to four weeks post-exposure, but the peak occurred in the second and the third weeks. Hair fell out in bunches on combing or gentle plucking. Onset correlated roughly with the exposure dose as estimated from distances and shielding. Epilation was an easily recognized sign following a latent period and sometimes clinically heralding the onset of radiation injuries. Survivors with early epilation usually had more severe syndromes. The extent of epilation did not correlate with prognosis. The scalp was the area most commonly involved by epilation; axillary and pubic hair were little affected. By the 12th to 14th weeks there was regrowth of hair; no permanent epilation was noted.

Purpura. Purpura was recognized as early as the third day and the peak was 20 to 30 days after the atomic bombs in both cities. As with epilation, the prevalence of purpura correlated closely with dose. A sharp drop in purpura was observed among survivors exposed beyond 1.5 kilometers (1 mile). The reported preva-

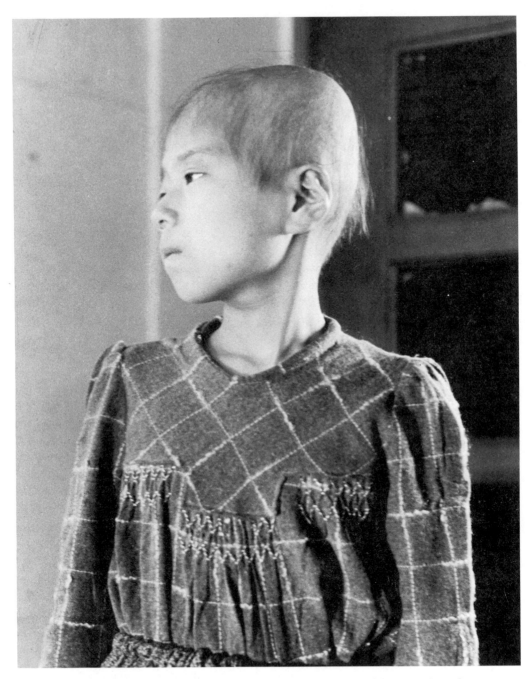

This girl has lost her hair. She was at home in a wooden house about 2 kilometers from the hypocenter. Hiroshima, October 6, 1945. (Shunkichi Kikuchi, Hiroshima–Nagasaki Publishing Committee.)

lence of purpura was undoubtedly lower than the actual because, if minimal, it frequently was not recorded.

Other Symptoms. Oropharyngeal lesions; hemorrhagic manifestations including nosebleed, uterine bleeding, and hemorrhages from the mouth, rectum, and urinary tract; and fever with or without apparent infection were undoubtedly signs related to radiation damage of the hematopoietic and gastrointestinal systems. Diarrhea was a common sign suggestive of radiation injury. However, there were many other causes for diarrhea because of conditions extant after the atomic bombs.

Outline of Clinical Course and Prognosis

It has been estimated that about 64,000 Hiroshima and 39,000 Nagasaki civilians were either killed immediately or died within about two months. Of these, about 50 percent died within 6 days and 96 percent died within 20 days. Since multiple injuries were common, the true causes of death were in many cases unknown, especially of those who died within three weeks. However, nearly all who died within three weeks had signs suggestive of radiation injuries. Furthermore, radiation injuries, with burns and other trauma, undoubtedly contributed to the total lethal effects.

The prodromal radiation syndrome was most marked in the very severely exposed, most dying within two weeks and having blood-cell abnormalities as well. Others dying 30 days after the bombs had milder prodromal syndromes.

In those less severely exposed, coincidental with the leukopenia and thrombocytopenia, signs of infections and hemorrhagic tendencies followed next in order and lasted from three to six weeks. The clinical courses and prognoses of survivors varied individually because of the diversity of combined thermal and blast injuries and various conditions of irradiation. The usual clinical course and severity of signs and symptoms observed in the first three weeks in each category are shown in Table 2, though there must have been considerable overlap. Dose-effect relations for humans are not precisely known, even for truly homogeneous whole-body exposure. However, comparison of published analyses with the categories of severity of exposure in Table 2 suggests exposure doses of 450 to 600 or more rads for Group I; 300 to 450

Table 2. *Atomic bomb survivors by clinical symptoms and signs of radiation injuries. (卌) (卄) (+) and (±) connote grade of symptoms and signs in order of decreasing severity.*

Degree of Severity*	First Week	Second Week	Third Week	Approximate Mortality and Time of Death
Very severe (Group I)	Nausea and vomiting (卌) Fever, apathy, delirium, diarrhea (卌) Oropharyngeal lesions (+)** Leukopenia (卌)	Fever (卄) Emaciation Leukopenia (卌) Anemia Hemorrhagic diathesis (±)[†] Epilation (±)		100% First and second weeks
Severe (Group II)	Nausea and vomiting (卄) Anorexia Fatigue	Fever (卄) Leukopenia (+) Anemia (±)	Anorexia, emaciation, fever, diarrhea, epilation (卌) Oropharyngeal lesions (卌) Hemorrhagic diathesis (卌) Leukopenia (卌) Anemia (卄)	50% Third to sixth weeks
Moderately severe (Group III)	Gastrointestinal[‡] syndrome (卄)	Leukopenia (+)	Anorexia, emaciation, fever, diarrhea, epilation (± ~卄) Oropharyngeal lesions (± ~卄) Hemorrhagic diathesis (± ~卄) Leukopenia (卄) Anemia (+)	Less than 10% Sixth week or later
Mild (Group IV)	Gastrointestinal syndrome (±)	Leukopenia (+)	Fever (±) Epilation (±) Oropharyngeal lesions (±) Hemorrhagic diathesis (±) Leukopenia (±)	None

*This classification is presented only as an orientation, and there must have been considerable overlap in each category. These descriptive terms were used by the Joint Commission for the Investigation of the Atomic Bomb in Japan.

**These lesions (ulcerations) occurred on all mucous membrane surfaces but were more prevalent in lymphoid areas. The tonsil, pharynx, larynx, nasal passages, and tongue were frequently involved.

[†]Hemorrhagic diathesis is the tendency to spontaneous bleeding or bleeding from a trivial trauma.

[‡]Gastrointestinal syndrome includes nausea, vomiting, anorexia, and diarrhea.

Table 3. Civilian casualties and mortality for Hiroshima and Nagasaki.
Population figures have been rounded off to the nearest thousand.

	Hiroshima	Nagasaki
Total casualties	136,000	64,000
Died first day	45,000	22,000
Died after first day	19,000*	17,000
Dead in four months	64,000**	39,000
Living injured first day	91,000	42,000
Surviving casualties	72,000	25,000
Uninjured	119,000	110,000

*Of this number, 86.5 percent died within 20 days.
**Of this number, 96.0 percent died within 20 days.
Source: Oughterson and Warren, 1956.

rads for Group II; 200 to 300 rads for Group III; and 100 to 200 rads for Group IV. Statistical analyses of correlation between mortality rates of survivors and distances from the hypocenters showed that 50 percent mortality among the lightly shielded was observed at 1.2 kilometers in both cities.

Numbers of Casualties

Hiroshima is built on the delta of the Ota River. The atomic bomb burst nearly over the center of the city. Nagasaki is located at the head of a long, narrow bay, is surrounded by hills, and extends along both sides of the bay and up into two valleys. The bomb exploded over one of these valleys, and the effects of the bomb were largely confined to this one valley. In Hiroshima, approximately 60 percent of the population was within 2000 meters of the hypocenter; in Nagasaki, only 30 percent was so situated. These differences influenced the effects in the two cities, loss of life and physical destruction having been greater in Hiroshima than in Nagasaki.

There is continuing controversy regarding the numbers of persons killed by these bombs. The numbers of civilian casualties estimated by the Joint Commission are shown in Table 3. These numbers do not include military casualties, nor survivors with late effects. No data concerning military casualties, which must have been large in Hiroshima, are available.

85

The numbers of persons among the survivors with burns, blast, or ionizing radiation injuries, or a combination thereof, were estimated by the Joint Commission. For Hiroshima, these victims numbered approximately 60,000 with burns, 78,000 with blast injuries, and 35,000 with radiation injuries. For Nagasaki, there were an esimated 41,000 burns, 45,000 blast injuries, and 22,000 radiation injuries. These may have occurred in combination.

Atomic Bomb Radiation in the Acute Stage

Background

The most conspicuous prodromal radiation syndrome includes nausea, vomiting, diarrhea, fever, hypotensive shock (shock associated with lowered blood pressure), and lassitude. The subsequent courses of those having the acute radiation syndrome is predictable by the severity and duration of the early symptoms and signs. This initial symptom complex has long been known as "radiation sickness," common in patients receiving radiation therapy, but it is only the initial syndrome. To avoid confusion with subsequent acute radiation symptoms such as hematopoietic depletion (a shutting down of the blood-forming system), the term "prodromal radiation syndrome" is more appropriate than "radiation sickness."

As to its severity, marked individual variability is involved. Recent studies indicate that with a sufficiently large single dose of deeply penetrating ionizing radiation, individual variability is minimized, and nearly all exposed individuals develop all prodromal signs. With a few thousand rads, all individuals can be expected to have the syndrome within 15 minutes, and it may persist for several days until their death.

The prodromal syndrome can be produced by irradiating the abdomen, thorax, or head. Epigastric (the upper middle area of the abdomen) irradiation elicits the response with the least dose, while irradiation of the extremities is not effective. Shielding the abdomen with lead during total body irradiation can protect against this response. These results also suggest that the autonomic nervous system plays an important role in the prodromal syndrome. Immediate diarrhea, fever, and hypotension seem to signify supralethal exposure.

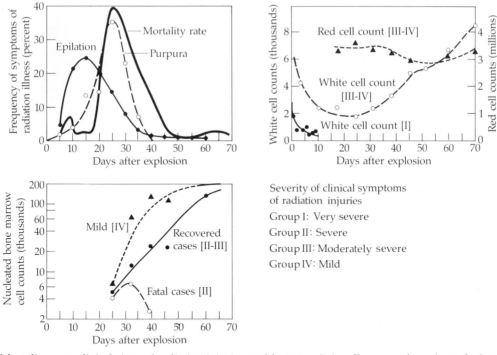

Mortality rate, clinical signs of radiation injuries, and hematopoietic cell counts of survivors during the first ten weeks after the Hiroshima atomic bomb.

Hematological Findings in the Acute Stage

Bone marrow depletion by irradiation is the most critical damage leading to death. The clinical course following atomic bomb radiation injuries, mortality, and some hematological data are shown in the figure above. Data in this figure were compiled from all sources, primarily in Hiroshima. The upper part of this figure shows the Hiroshima mortality curve to be bimodal, the first small peak having occurred eight or nine days after exposure. By historical classification, the term "gastrointestinal death" due to total body irradiation was at times proposed for these early deaths. These victims who died early had extensive bone marrow damage, manifested by leukopenia and thrombocytopenia. However, "gastrointestinal death" is now suggested to be a rapid form of hematopoietic death rather than as originally assumed, "death resulting from gastrointestinal damage."

The second mortality peak occurred between the 20th and 40th days in both cities. The death rates of survivors diminished markedly thereafter. The figure shows that the frequency of purpura correlated with the second mortality peak. Fever (not shown in the figure) was highest in frequency between the 20th and 30th days. Conversely, counts of nucleated cells of the bone marrow and of circulating white blood cells were profoundly reduced during the same period. Epilation was a first warning symptom of the second critical phase.

Results clearly indicate that the sequence of events in heavily irradiated people was as follows: severe, persistent bone marrow depletion causing a decrease in the number of all blood cells, leading to infections, oropharyngeal lesions, fever, a hemorrhagic tendency, and death. Denuding of the villi (projections that serve as sites for absorption of nutrients) of the intestinal epithelium may play a significant role.

In spite of early recognition of the syndrome of radiation injury by Japanese doctors, few specimens of bone marrow were obtained during the first three weeks because the attending physicians were too busy to perform biopsies. Only two bone marrow specimens, the earliest, were obtained on the sixth day, from Hiroshima survivors exposed in a concrete building 250 meters from the hypocenter. Their nucleated cell counts were not recorded, but drastic reductions in numbers of bone marrow cells were apparent. Both patients had slight burns and their circulating white blood cells numbered only $300/mm^3$ (normal values are 5000–$10,000/mm^3$). Their platelet counts were about one-third of normal and they died on the eighth and eleventh post-exposure days.

After the arrival of research teams from medical schools about four weeks after the bombs, physicians began recording bone marrow findings. The lowest portion of the figure on p. 87 shows the mean values of nucleated bone marrow cell counts demonstrating the regenerating processes of the bone marrow with time. At the fourth week, the average bone marrow cell count was very low in each group. Subsequently, among the cases in Group II who died within six weeks, there was no recovery until death. In cases who recovered (Groups II, III, and IV), low values were still evident at the seventh week, but gradual regenerating processes were seen after that, and their bone marrow cell counts reached normal levels. In those very severely injured cases, low white blood cell counts progressed rapidly and counts were usually less than 500/mm^3 between the fifth and twelfth days. In those who survived more than ten days, the greatest white blood cell count depression occurred between three and five weeks post-exposure. Depression of white blood cell counts from the third to fifth week was found to correlate with death and was concluded to be the best prognostic indicator.

A number of factors obscured the red blood cell counts of atomic bomb survivors, especially blood loss from wounds, dehydration by severe burns and by diarrhea, bone marrow depletion, hemorrhagic tendencies, and severe infections. In severely exposed cases (Group II), the red cell counts declined at a steady rate and their minimum values were reached at about the same time as those of white cell counts, the minimum of which, in many cases, coincided with death. There is no clear evidence, however, that the degree of anemia is useful in prognosis. In the moderately and mildly exposed cases (Groups III, IV), the lowest red cell counts and hemoglobin concentrations were observed from six to nine weeks after the bombs, and their mean values had not returned to normal, even by the twelfth week. This was supported by the low counts of regenerated young red blood cells that were obtained between eight and twelve weeks. More than 60 percent of these were less than normal.

Platelet counts soon after the bombing were recorded for only two of the very severely exposed Hiroshima cases, both of them six days after the bomb. The total platelets were 87,400/mm^3 and 64,500/mm^3 (normal counts: 200,000–500,000/mm^3), and the patients died on the eighth and ninth days. Neither had skin purpura. Of 20 patients in Group I and dying within ten days, only 1 had cutaneous purpura. Purpura of the organs was frequently noted, however, in autopsy cases during the same period. In Groups II and III, the frequency of platelet counts of less than 25,000/mm^3 was found in 83.3 percent of the fatal and 20.8 percent of the nonfatal cases. Thus, the severity of thrombocytopenia correlated directly with radiation dose and afforded a rough index of survival.

Laboratory data on hemostasis (the arrest of bleeding) and coagulation in the survivors were scarce, and none were obtained for those who received the highest doses because all of them died within two weeks of the bombs. In the severely exposed (Group

II), moderate prolongation of whole blood coagulation time was reported. Prolonged bleeding times were concomitant with the onset of hemorrhagic diathesis (tendency to spontaneous bleeding or bleeding from a trivial trauma), and their degrees were proportional to the decreases in platelet counts. As platelet counts approached normal levels, bleeding times became correspondingly shorter, and all had returned to normal by the ninth week post-exposure.

Recent studies have shown that, after the platelets, fibrinogen (a blood constituent critical to clot formation) is the next most radiosensitive clotting factor. Low coagulability with deficient clot formation, caused by qualitative changes in fibrinogen, is suspected to have occurred in the severely irradiated survivors. Increased capillary fragility is also responsible for post-radiation hemorrhage.

Space does not permit a detailed description of recovery of hematopoietic cells. In most survivors, blood values had reached normal by the end of the second year, but many instances of various blood abnormalities (i.e., anemia, eosinophilia (an abnormally large number of eosinophils—a type of white blood cell), leukopenia, thrombocytopenia, and capillary fragility) have all been reported, especially among the heavily irradiated survivors eight to eleven years after the atomic bombs.

Relative Biological Effectiveness of Atomic Bomb Neutrons

Mortality from ionizing radiation among survivors who were shielded, mainly by Japanese-style houses, was estimated by the Joint Commission using a population sampling method. The curves obtained reveal that mortality was 50 percent at a distance of 1.2 kilometers ($\frac{3}{4}$ mile) from the hypocenter in each city. The ratios of gamma to neutron dose to the survivors were very different in the two cities. According to the Atomic Bomb Casualty Commission estimates for air-dose values, in Hiroshima the radiation dose at 1.2 kilometers was about 154 rads, consisting of 95 rads of gamma rays and 59 rads of neutrons. In Nagasaki, the air-dose value at 1.2 kilometers was about 403 rads: 392 rads of gamma rays and 11 rads of neutrons. The Nagasaki neutron dose was so small that, for practical purposes, survivors at this distance can be considered to have received only gamma radiation. Thus, the relative biological effectiveness (RBE) of atomic bomb neutrons at 1.2 kilometers for acute lethality can be calculated:

$$95 + (59 \times \text{RBE}) = 392$$
$$\text{RBE} \approx 5$$

Effects on Spermatogenesis

From 1.5 to 3 months after the explosion, semen of 131 Hiroshima atomic bomb survivors were examined. No abnormalities were found in the physical or chemical character of seminal fluid, indicating normal function of the prostate gland. However, good correlation between sperm count and exposure distance was noted. Sperm counts of less than 10×10^6/cc semen were found mostly in proximally exposed survivors. (Normal sperm count is 20–60×10^6/cc; many authorities consider 20×10^6/cc the lower limit of fertility.)

Low sperm counts were found in 42 (70 percent) of 60 persons exposed within 1.5 kilometers and in 18 (25 percent) of 71 exposed beyond this distance. Supportive of this were autopsy findings that as early as the fourth day after exposure, the testes of victims showed extensive death of the sperm germ cells. These were most marked in those who died early and had been within 1.5 kilometers of the hypocenter. High frequencies of morphological abnormalities and reduced sperm motility were also recorded in proximally exposed survivors. Nine months after exposure, sperm counts in 11 of 12 survivors exposed within 1.6 kilometers were less than 10×10^6/cc. Twenty-two months after the bomb, low sperm counts were still evident in 11 (41 percent) of 27 survivors exposed within 1.6 kilometers.

The next sperm counts in January 1959, 13.5 years after the bomb, confirmed 3 Hiroshima cases having an absence of sperm in seminal fluid. They were exposed within 1.3 kilometers from the hypocenter, and each had a clinical history of radiation injury in the acute stage. However, sperm counts of 10 survivors exposed within 1.6 kilometers and reexamined at this same time had all returned to normal. There was recovery of sperm motility and reduced numbers of abnormal forms.

Results obtained from study of nuclear accidents and the Bikini survivors show the same pattern of sperm count fluctuations. The greatest depression was observed between the seventh and ninth months after exposure, and recovery, when possible, usually occurred by the end of the second year.

References

Collection of the Reports on the Investigation of the Atomic Bomb Casualties. 2 vol. 1953. Japan Society for the Promotion of Science. Tokyo: Maruzen.

Guskova, A. K., and G. D. Baysogolov. 1971. "Radiation Sickness in Man." Translation series, AEC-tr-7401.

Kajitani, T., and S. Hatano. 1953. "Medical Survey on Acute Effects of the Atomic Bomb in Hiroshima." In *Collection of the Reports of the Investigation of the Atomic Bomb Casualties,* vol. 1. Japan Society for the Promotion of Science. Tokyo: Maruzen.

Oughterson, A. W., G. V. LeRoy, A. A. Liebow, E. C. Hammon, H. L. Bernett, J. D. Rosenbaum, and B. A. Schneider. 1951. *Medical Effects of Atomic Bombs: The Report of the Joint Commission for the Investigation of the Atomic Bomb in Japan.* 6 vol. Washington, D.C.: U.S. Atomic Energy Commission.

Oughterson, A. W., and S. Warren, eds. 1956. *Medical Effects of the Atomic Bomb in Japan.* New York: McGraw-Hill.

Warren, S., and J. Z. Bowers. 1950. "The Acute Radiation Syndrome in Man." *Annals of Internal Medicine,* 32:207–216.

8

Delayed Medical Effects
at Hiroshima and Nagasaki

Takeshi Ohkita, M.D.

Soon after the end of World War II, the Japanese and United States governments sent teams to investigate the effects of the Hiroshima and Nagasaki atomic bombs. Although it was known in 1945 that ionizing radiation could induce mutations, cancers, and other deleterious effects in plants and experimental animals, little was known about how ionizing radiation from atomic bombs might affect the exposed survivors in the succeeding years.

The Atomic Bomb Casualty Commission (ABCC) was organized in 1946 under the direction of the National Research Council of the National Academy of Sciences in the United States, and in 1948 the National Institute of Health of the Japanese Ministry of Health and Welfare formally joined in its studies. The Research Institute for Nuclear Medicine and Biology was started at Hiroshima University in 1961, and the Atomic Diseases Institute at Nagasaki University in 1962. In 1975, the ABCC was dissolved and the Radiation Effects Research Foundation (RERF) replaced it as a joint enterprise of the Japanese and American governments, to continue the surveillance of the Hiroshima and Nagasaki atomic bomb survivors for long-term aftereffects.

In this report, a brief summary will be presented on the major points based on careful and painstaking studies carried out by many American and Japanese scientists. Table 1 gives the

Table 1.   Main studies on the aftereffects of the atomic bomb, Hiroshima and Nagasaki.

Keloids and overgrowth of scars	Malignant tumors
Blood disorders	Chromosome changes
Eye lesions	Genetic effects
Disturbances of reproductive function	Aging and life span
Effects on fetuses exposed in utero	Psychoneurological disorders
Growth and developmental disturbances	

major research studies made to date on the late effects of the atomic bomb.

Keloids and Overgrowth of Scars

After the healing of some very severe burns due to the radiant heat from the bombs in both Hiroshima and Nagasaki, overgrowth of scar tissue and keloids were frequently observed among those who were burned within 1.7 kilometers (1 mile) of the hypocenters. The incidence of keloids on various parts of the body in 1946–1947 varied between 20 and 50 percent for thermal burns studied. Formation of keloids occurred mostly on exposed parts of the body such as face, neck, and extremities. As keloids are physically deforming, are accompanied by stinging pain and itching, and are associated with functional disorders of the joints, survivors suffered both physically and mentally.

Surgeons in both Hiroshima and Nagasaki devoted a great deal of effort to the treatment of keloids, however, they often recurred after surgical removal or skin transplantation. Many keloids transformed to ordinary scars during the 10-year period following 1946 and 1947, the peak years for keloid formation.

Blood Disorders

After the bombing the blood cells of survivors were extensively investigated. These investigations have continued uninterrupted over the years. In most survivors, the blood cell values, which

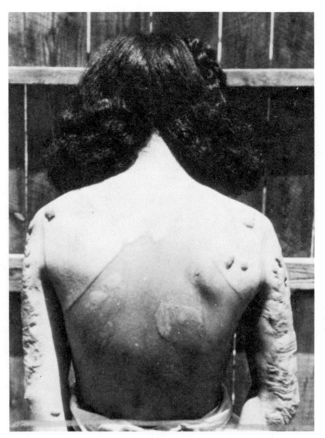

Keloid formation on burned skin. This girl was exposed at about 1100 meters from the hypocenter in Hiroshima, July 1947. (Courtesy of Dr. T. Ohkita.)

frequently had significant abnormalities, returned to the normal range by the end of the second year. Delayed blood disorders, however, were more insidious and required careful study during the last 35 years for their possible cause-and-effect relationship to atomic exposure. These are summarized below:

• The incidence of leukemia among the atomic bomb survivors, which had increased significantly during the mid-1950's, has decreased steadily since then, but the leukemia risk in heavily exposed survivors, especially in Hiroshima, is still higher than the average value for Japan. No relationship has been demonstrated between the incidence of leukemia among children exposed in utero and nonexposed children born later to atomic bomb survivors.

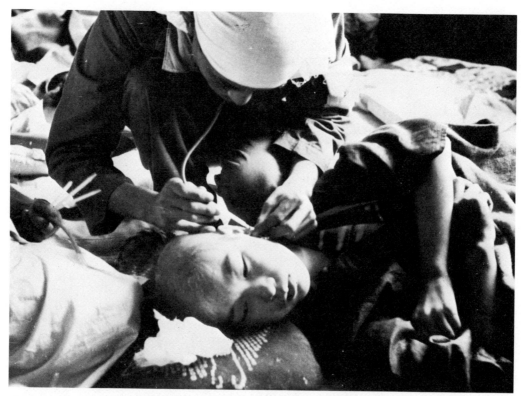

A nurse taking a blood sample from this girl's ear. The girl has lost her hair, an early symptom of radiation syndrome. Nagasaki, early September 1945. (Yasuo Tomishige, Hiroshima–Nagasaki Publishing Committee.)

- The incidence of multiple myeloma has been found to be significantly elevated 20 years after the bombing in the heavily exposed survivors.

- The risk of developing malignant tumors in lymphoid tissue has been found to be linked to radiation dose, but the dose relationship is not conclusive.

- Cases of idiopathic myelofibrosis have been reported among atomic bomb survivors in autopsy subjects, but it is not possible to assert that this is due to radiation, because of the complexities of the disease pattern, the paucity of cases, and the inconclusiveness of the link between autopsy findings and exposure dose.

- The incidence of aplastic anemia and polycythemia vera has been shown clearly not to be related to atomic bomb exposure.

Eye Lesions

Following exposure to the atomic bomb, eye lesions were numerous and of three types: (1) lesions from direct injury (thermal burns especially of the eyelids; trauma due to foreign bodies, especially glass splinters; and flash lesions to the cornea, conjunctiva, and retina), (2) lesions from the effects of radiation illness, including those from anemia, hemorrhage, and infection, and (3) delayed effects, including cataracts and scar deformation.

Among the delayed effects, atomic bomb cataracts were the first

---

anaplastic carcinoma   cancer in which there is loss of structural differentiation in cells.

aplastic anemia   anemia caused by deficient production of red blood cells.

granulocytic leukemia   leukemia of the white blood cells that contain granules.

idiopathic myelofibrosis   replacement of the bone marrow by fibrous tissue; idiopathic means of unknown origin.

keloid   scar formation characterized by irregular shape, raised sharply above the surrounding skin.

leukemia   progressive malignant disease of the blood-forming organs characterized by a proliferation of white blood cells in the blood and bone marrow. In acute leukemia death occurs within a few months of the onset of symptoms; the duration of chronic leukemia exceeds one year, with a gradual onset of symptoms.

lymphocytic leukemia   leukemia of the lymphocytes, white blood cells that do not contain granules and are formed in lymph nodes and bone marrow.

multiple myeloma   progressive, usually fatal, malignant tumor characterized by an infiltration of bone and bone marrow and accompanied by anemia and kidney lesions.

polycythemia vera   chronic, proliferative disorder of all bone marrow elements, resulting in an increased red blood cell mass.

rad   unit of radiation; 450 rads in a short period of time is a lethal dose for approximately 50 percent of healthy adults exposed.

---

and the most frequent to be noted (by H. Ikui in 1948 (and 1967) in Hiroshima, and by K. Hirose and S. Fujino in 1949 (and 1950) in Nagasaki). Later studies (I. Hirose and A. Okamoto, 1961) revealed that the frequency of cataracts increased with the estimated exposure dose. The incidence, for example, for doses exceeding 100 rads was 55 percent. (Studies were done to control for the interference of senile cataracts in the data.)

Disturbances of Reproductive Function

Disturbances of reproductive function are an inevitable consequence of exposure to atomic radiation. Loss of sexual desire was a frequent occurrence during the two to three months after the explosion. Investigations of the sperm of those exposed in Hiroshima were carried out by the Tokyo University Team. Of 124 cases studied in 1945, approximately one-third had markedly reduced sperm counts (less than 5×10^6/cc), indicating sterility.

According to the results of a survey performed in 1946 and 1947, there were still considerable numbers of cases of sterility among those men exposed to the bomb within 1.5 kilometers from the hypocenter. Later surveys indicated that the majority returned to almost normal in five years. According to autopsy data gathered between 1951 and 1963, testicular tissue showed greater evidence of age-related degenerative change among exposed males than among age-matched nonexposed controls.

Among women, the most prominent symptom appearing soon after the explosion was menstrual disorder. Seventy-two percent of 500 women studied showed menstrual disorders after exposure. The incidence of abnormal menstruation correlated with the distance from the hypocenter. Menstrual disturbance occurred less with thermal injury and trauma. Approximately 78 percent of those women surveyed with abnormal menstruation returned to normal by March 1946. A later survey by the Atomic Bomb Casualty Commission failed to demonstrate any evidence of reduced fertility in individuals exposed during early adulthood, the prepubertal period, or in utero.

Effects on Fetuses Exposed In Utero

Fetuses in utero were also affected by the atomic bombings. Many pregnant women died; others experienced fetal death or abortion.

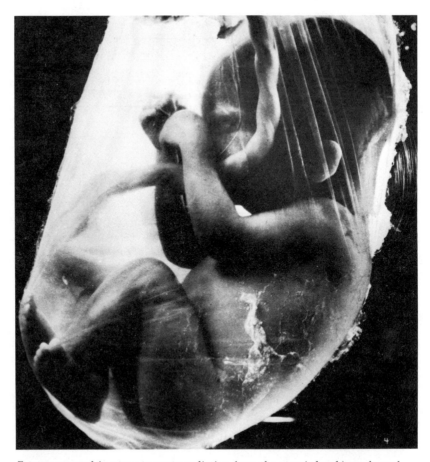

Fetuses exposed in utero to acute radiation from the atomic bombings showed a marked increase in fetal and infant mortality and in mental retardation. (United Press International.)

Those babies born after a normal pregnancy were referred to as "children exposed in utero," although no accurate data exist on the exact number of those exposed in utero at the time of the bombings. Figures gathered by the Atomic Bomb Casualty Commission in 1960 revealed 2310 Hiroshima survivors who had been exposed in utero and 1562 for Nagasaki.

J. N. Yamazaki and his co-workers studied 98 women who had been pregnant on August 9, 1945, within 2 kilometers ($1\frac{1}{4}$ miles) of the hypocenter in Nagasaki. One hundred and thirteen women who had been exposed at a distance of 4 to 5 kilometers of the hypocenter were used as controls. Tables 2, 3, and 4 show the results of their survey on the incidence of fetal deaths, the mortal-

Table 2. Fetal mortality among irradiated and control groups.

Distance from Hypocenter (km)	Group	Number of Conceptions	Number of Abortions	Number of Stillbirths	Fetal Mortality (%)
0–2.0	Radiation signs	30	3	4	23.3
0–2.0	No radiation signs	68	1	2	4.4
4.0–5.0	Controls	113	2	1	2.7

Source: Yamazaki et al., 1954.

ity rate of newborns and infants, and the morbidity rate of the children up to five years after birth.

Among the 30 mothers who had been exposed within 2 kilometers of the hypocenter and who showed acute radiation symptoms, there were 7 fetal deaths, 6 neonatal and infantile deaths, and mental retardation in 4 out of the remaining cases, with an overall morbidity and mortality of 60 percent. The corresponding morbidity and mortality rates for the mothers exposed within 2 kilometers but without radiation symptoms was 10 percent and for the control mothers, 6 percent.

Microcephaly or small head size (defined as a head circumference less than two standard deviations below the age- and sex-specific mean head size) is one of the most regrettable aftereffects of atomic bomb radiation. Children with this condition were usually mentally retarded, with an extremely low intelligence quotient (IQ 16–25) and very low social adaptability. R. W. Miller and W. J. Blot (1972) studied the relation between microcephaly and radiation exposure dose. Forty-eight persons with microcephaly were found among those exposed in utero in Hiroshima, and 15 in Nagasaki. The highest incidence of microcephaly, especially when accompanied by mental retardation, was encountered in those who had been exposed before the 18th week of fetal life especially from the 3rd to the 15th weeks. In Hiroshima, a significant increase in the frequency of microcephaly was already seen among those whose mothers had received low doses of radiation, such as 10 to 19 rads.

100

Table 3. Neonatal and infant mortality among irradiated and control groups.

Distance from Hypocenter (km)	Group	Mothers*	Neonatal Deaths	Infant Deaths	Mortality (%)
0–2.0	Radiation signs	23	3	3	26.1
0–2.0	No radiation signs	65	3	0	4.6
4.0–5.0	Controls	110	1	3	3.6

*Mothers of living infants; stillbirths and abortions are excluded.
Source: Yamazaki et al., 1954.

Table 4. Child morbidity among irradiated and control groups.

Distance from Hypocenter (km)	Group	Mothers*	Mental Retardation in Child	Child Alive and Normal After One Year of Life	Rate of Mental Retardation (%)
0–2.0	Radiation signs	16	4	12	25
0–2.0	No radiation signs	60	1	59	1.6
4.0–5.0	Controls	106	0	106	0

*Mothers whose children were alive at time of examination.
Source: Yamazaki et al., 1954.

In general, the frequency of microcephaly increased with the increase of exposure dosage. On the other hand, in Nagasaki, no mentally retarded microcephalic children were observed under 150 rads. This difference in frequency of microcephaly with mental retardation between the two cities might be attributable to the difference in radiation quality between Hiroshima and Nagasaki, that is, to the greater proportion of neutron radiation in Hiroshima.

Growth and Developmental Disturbances

Studies performed on survivors directly exposed in childhood during the decade following the explosions suggest a detrimental effect on growth. Between 1966 and 1968, height and weight measurements were obtained by the ABCC for approximately 3200 people who were under 18 when the bomb was dropped on Hiroshima. Average heights were lower among those exposed to more than 100 rads in early childhood. The differences among the dose groups diminished with increasing age. For example, those who were under 6 at the time of the Hiroshima bomb, who were exposed to more than 100 rads, had the greatest reductions in average adult height. Children who were 6 to 11 years old with more than 100 rads exposure still had reduced adult height, but to a lesser degree. No apparent difference in mean height between the dose groups was observed for those aged 12 to 17 years at the time of the bomb. These findings were true for both Hiroshima males and females.

In Nagasaki, by contrast, a significant dose effect was not observed. Smaller heights were seen in Nagasaki females exposed to 100 or more rads when less than 5 years old at the time of the bomb, but not in Nagasaki males. The difference in effect between the two cities was again thought possibly attributable to the differing quality of radiation, however, non-radiation-related factors could not be discounted.

Malignant Tumors

The most significant radiation effect has been the induction of malignant tumors in exposed survivors. The earliest evidence of radiation-induced malignant change was the occurrence of increased leukemia in the late 1940's and early 1950's. The figure on p. 103 shows the number of leukemia cases observed annually among Hiroshima survivors exposed within 2 kilometers of the hypocenter up to 1978. The largest number was observed between 1950 and 1953 with a peak in 1951.

According to the joint studies conducted by the RERF and the Hiroshima and Nagasaki Universities Medical Schools during the period from 1950 to 1978, 202 leukemia cases were identified in a fixed group of atomic bomb survivors and their controls. The

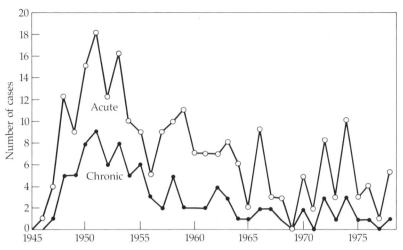

Number of leukemia cases among Hiroshima survivors exposed within 2000 meters of the hypocenter, according to year of onset. (Courtesy of Dr. T. Ohkita, 1981.)

analysis showed that the risk for all types of leukemia, excluding chronic lymphocytic type, increased with radiation dose in both cities, except for those survivors who received less than 100 rads in Nagasaki. The risk at every dose level has been greater in Hiroshima than in Nagasaki. The incidence of chronic granulocytic leukemia increased in Hiroshima among survivors who received less than 50 rads, but in Nagasaki it was elevated significantly only among survivors who received 200 rads or more.

The leukemia-causing effect of radiation was different in relation to age at exposure. The overall incidence rate of leukemia among survivors who were exposed to more than 100 rads in Hiroshima was about 60 per 100,000 persons, about 15 times greater than that in the controls.

The occurrence of solid tumors in survivors generally increased from around 1960, after the peak of leukemia incidence. In the early stage of studies until 1968, a definite correlation between exposure distance or dose and malignancy incidence was reported for cancers of thyroid, breast, and lung.

Clinical studies in both cities in the late 1950's and early 1960's showed that the frequency of thyroid cancer was higher among survivors, especially women who were exposed to high radiation dose, than that in the nonexposed. According to the survey made by L. N. Parker and his co-workers (1977) during the period from 1958 to 1971, the relative risk for thyroid cancer for survivors exposed to 50 rads or more was about 2.5 times that of the controls. Aside from clinically evident thyroid cancers, pathological investigations have revealed large numbers of malignant tumors of the thyroid that did not produce symptoms. Of 3067 cases autopsied up to 1968, 536 primary thyroid cancers were found, 98 percent of which were latent papillary adenocarcinomas (a specific type of thyroid cancer). The relative risk of latent thyroid cancer for those exposed to more than 50 rads was 1.4 times higher than for the nonexposed. A subsequent study of thyroid cancer after 1971 is in progress.

The death surveys made in Hiroshima about ten years after the bombing all showed that mortality from breast cancer among the exposed was higher than among the nonexposed and than the national average. M. Tokunga and his colleagues (1977) reported that during the period from 1950 to 1974, the relative risk for women exposed to 100 rads or more was about 3.3 times that of the controls. The risk was higher in heavily exposed women who were 10–39 years old at the time of the bombing, but not among those who were between the ages of 40 and 49. These differences in age susceptibility to radiation-related breast cancer suggest that hormonal factors might be involved. The dose response relationships in both cities appear to be linear and are of similar magnitude. No variation in the mean time from exposure to the appearance of breast cancer with exposure dose was noted. The histological type of breast cancer did not differ by exposure dose.

The first case of lung cancer among atomic bomb survivors in Hiroshima was reported in 1953. The radiation effect of the bombing on the lung was definite by 1958. Based on the survey up to 1974, G. W. Beebe and H. Kato (1975) reported that the relative risk for survivors exposed to 100 rads or more, compared with the nonexposed and those exposed to less than 10 rads, was 1.8 times. According to the pathological study by R. W. Cihak and co-workers (1974), small cell anaplastic carcinomas were definitely increased in heavily irradiated persons compared to the controls.

The most recent and reliable information on cancer mortality with atomic bomb radiation can be obtained from the Life Span Study, which has been in progress since 1950 by the ABCC and RERF. In addition to the thyroid, breast, and lung, people exposed to 200 rads and more demonstrated a significantly increased risk

Table 5. *Bone marrow chromosome aberration in Hiroshima atomic bomb survivors.*

Distance from Hypocenter (km)	Number of Cases Examined	Number of Cells Observed	With Chromosome Aberration	
			Number of Cells (%)	Number of Cases (%)
< 0.5	20	1127	262 (23.2)	18 (90.0)
0.6–1.0	21	789	101 (12.8)	11 (52.4)
1.1–1.5	18	556	1 (0.2)	1 (6.2)
1.6–2.0	23	728	0	0
2.1–3.0	23	737	3 (0.4)	1 (4.3)
Control	17	624	0	0

Source: Kamada et al., 1979.

for developing cancers of the esophagus, stomach, colon, and urinary tract. An increased risk to persons exposed to heavy radiation for the development of malignant salivary gland tumors and primary brain tumors in males has also been suggested.

Chromosome Changes

Chromosome aberration in atomic bomb survivors was initially reported by T. Ishihara and T. Kumatori in 1965. Since then, many studies have revealed that chromosomal aberrations in the peripheral blood lymphocytes (white blood cells that participate in immunity), bone marrow cells, and fibroblasts (immature fiber-producing cells of connective tissue) increased significantly in survivors who were exposed to high doses of radiation whether in utero or after birth.

The frequency of aberrant cells and of chromosomal aberrations per cell is closely associated with the increase in radiation dose, as shown in Table 5 for Hiroshima survivors.* These aberrations are all of the stable type. It is clear that the aberrations are the consequence of radiation injury and that the casualties from radiation exposure at a cellular level have not yet healed. At the present

*The frequency of aberrant cells at all dose levels is higher in Hiroshima than in Nagasaki (A. A. Awa, 1975).

time, almost all the survivors with chromosome abnormalities have normal blood values and are in good health, but the biological implication of chromosomal aberrations of somatic cells for the health of the exposed is still unknown.

Genetic Effects

Based on the experimental data, there was great concern about possible genetic effects in the offspring of survivors exposed to atomic bomb radiation in Hiroshima and Nagasaki. Careful genetic surveys have thus been carried out over the years since 1948. These studies looked at the frequency of gene mutations and chromosomal aberrations in children born to radiation-exposed parents.

In summary, genetic surveys undertaken to date have yielded no positive evidence for a genetic hazard due to atomic bomb radiation. It is possible that genetic effects do not show up so distinctly in human beings, since they have fewer pregnancies as well as fewer fetuses per pregnancy than do other mammals. Furthermore, human populations are very heterogeneous, and the chances of mutation, which would almost always be the result of having an offspring receive an identical mutant gene from each parent, would be extremely low. Thirty-seven years—only two generations—have passed since the explosion of the atomic bombs. This is a very short interval for human genetic effects. It is still too early to say definitively that there has been no genetic effect from the atomic bombs at Hiroshima and Nagasaki.

Conclusion

I believe that the pledge not to repeat the mistakes of Hiroshima and Nagasaki can be made a lasting reality if the people of the world realize and understand the suffering of those under the mushroom clouds in the two cities.

References

Hiroshima and Nagasaki: The Physical, Medical, and Social Effects of the Atomic Bombings. 1981. Edited by the Committee for the Compilation of Materials on Damage Caused by the Atomic Bombs in Hiroshima and Nagasaki. New York: Basic Books.

Kamada, N., A. Kuramoto, T. Katsuki, and Y. Hinuma. 1979. "Chromosome Aberrations in B Lymphocytes of Atomic Bomb Survivors." *Blood, 53*:1140.

Yamazaki, J. N., S. W. Wright, and P. M. Wright. 1954. "Outcome of Pregnancy in Women Exposed to the Atomic Bomb in Nagasaki." *American Journal of Diseases of Children, 87*:448.

References cited in the article that are not listed above can be found in *Hiroshima and Nagasaki: The Physical, Medical, and Social Effects of the Atomic Bombings.*

SECTION III

Nuclear War, 1980's: The Physical and Medical Consequences

There is no sensible military use for any of our nuclear forces: intercontinental, theater or tactical.

Admiral Noel Gayler, U.S. Navy (Ret.)
Former Director of the National Security Agency,
Commander of All U.S. Forces in the Pacific,
and Deputy Director of the Joint Strategic Target Planning Staff
The Washington Post, June 23, 1981

9

The Effects of Nuclear War

Office of Technology Assessment,
United States Congress

Nuclear war is not a comfortable subject. Throughout all the variations, possibilities, and uncertainties that this study describes, one theme is constant—a nuclear war would be a catastrophe. A militarily plausible nuclear attack, even "limited," could be expected to kill people and to inflict economic damage on a scale unprecedented in American experience; a large-scale nuclear exchange would be a calamity unprecedented in human history. The mind recoils from the effort to foresee the details of such a calamity, and from the careful explanation of the unavoidable uncertainties as to whether people would die from blast damage, from fallout radiation, or from starvation during the following winter. But the fact remains that nuclear war is possible, and the possibility of nuclear war has formed part of the foundation of international politics, and of U.S. policy, ever since nuclear weapons were used in 1945.

The premise of this study is that those who deal with the large issues of world politics should understand what is known, and perhaps more importantly what is not known, about the likely

Excerpted from *The Effects of Nuclear War* by the Office of Technology Assessment, U.S. Congress, 1979. The figures on pages 122, 124, and 133 are in the OTA report; the remainder are from other sources.

It is time to recognize that no one has ever succeeded in advancing any persuasive reason to believe that any use of nuclear weapons, even on the smallest scale, could reliably be expected to remain limited.

McGeorge Bundy, George F. Kennan,
Robert S. McNamara, and Gerard Smith
"Nuclear Weapons and the Atlantic Alliance,"
Foreign Affairs, Spring 1982

When a war starts, every nation will ultimately use whatever weapon has been available. That is the lesson learned time and again.

Admiral Hyman G. Rickover
Farewell Testimony to the Joint Economic
Committee of the U.S. Congress
Washington, D.C., February 8, 1982

---

consequences if efforts to deter and avoid nuclear war should fail. Those who deal with policy issues regarding nuclear weapons should know what such weapons can do, and the extent of the uncertainties about what such weapons might do.

Findings

The effects of a nuclear war that cannot be calculated are at least as important as those for which calculations are attempted. Moreover, even these limited calculations are subject to very large uncertainties.

Conservative military planners tend to base their calculations on factors that can be either controlled or predicted and to make pessimistic assumptions where control or prediction is impossible. For example, planning for strategic nuclear warfare looks at the extent to which civilian targets will be destroyed by blast and discounts the additional damage which may be caused by fires that the blast could ignite. This is not because fires are unlikely to cause damage, but because the extent of fire damage depends on factors such as weather and details of building construction that make it much more difficult to predict than blast damage. While it is proper for a military plan to provide for the destruction of key targets by the surest means even in unfavorable circumstances, the nonmilitary observer should remember that actual damage is likely to be greater than that reflected in the military calculations. This is particularly true for indirect effects such as deaths resulting from injuries and the unavailability of medical care, or for economic damage resulting from disruption and disorganization rather than from direct destruction.

For more than a decade, the declared policy of the United States has given prominence to a concept of "assured destruction": the capabilities of U.S. nuclear weapons have been described in terms of the level of damage they can surely inflict even in the most unfavorable circumstances. It should be understood that in the event of an actual nuclear war, the destruction resulting from an all-out nuclear attack would probably be far greater. In addition to the tens of millions of deaths during the days and weeks after the attack, there would probably be further millions (perhaps further tens of millions) of deaths in the ensuing months or years. In addition to the enormous economic destruction caused by the actual nuclear explosions, there would be some years during which the residual economy would decline further, as stocks were consumed and machines wore out faster than recovered production could replace them. Nobody knows how to estimate the likelihood that industrial civilization might collapse in the areas attacked; additionally, the possibility of significant long-term ecological damage cannot be excluded.

The situation in which the survivors of a nuclear attack find themselves will be quite unprecedented. The surviving nation would be far weaker—economically, socially, and politically— than one would calculate by adding up the surviving economic assets and the numbers and skills of the surviving people. Natural resources would be destroyed; surviving equipment would be designed to use materials and skills that might no longer exist; and indeed some regions might be almost uninhabitable. Furthermore, pre-war patterns of behavior would surely change, though in unpredictable ways. Finally, the entire society would suffer from the enormous psychological shock of having discovered the extent of its vulnerability.

From an economic point of view, and possibly from a political and social viewpoint as well, conditions after an attack would get

Hiroshima one year after the atomic bomb. Despite massive aid from the outside, the city remains largely in ruins. (New York Times.)

worse before they started to get better. For a period of time people could live off supplies (and, in a sense, off habits) left over from before the war. But shortages and uncertainties would get worse. The survivors would find themselves in a race to achieve viability (i.e., production at least equaling consumption plus depreciation) before stocks ran out completely. A failure to achieve viability, or even a slow recovery, would result in many additional deaths, and much additional economic, political, and social deterioration. This post-war damage could be as devastating as the damage from the actual nuclear explosions.

Uncertainties

There are enormous uncertainties and imponderables involved in any effort to assess the effects of a nuclear war, and an effort to look at the entire range of effects compounds them. Many of these

114

uncertainties are obvious ones: if the course of a snowstorm cannot be predicted one day ahead in peacetime, one must certainly be cautious about predictions of the pattern of radioactive fallout on some unknown future day. Similar complexities exist for human institutions: there is great difficulty in predicting the peacetime course of the U.S. economy, and predicting its course after a nuclear war is a good deal more difficult. This study highlights the importance of three categories of uncertainties:

- Uncertainties in calculations of deaths and of direct economic damage resulting from the need to make assumptions about such matters as time of day, time of year, wind, weather, size of bombs, exact location of detonations, location of people, availability and quality of sheltering.
- Effects that would surely take place, but whose magnitude cannot be calculated. These include the effects of fires, the shortfalls in medical care and housing, the extent to which economic and social disruption would magnify the effects of direct economic damage, the extent of bottlenecks and synergistic effects, the extent of disease.
- Effects that are possible, but whose likelihood is as incalculable as their magnitude. These include the possibility of a long downward economic spiral before viability is attained, the possibility of political disintegration (anarchy or regionalization), the possibility of major epidemics, and the possibility of irreversible ecological changes.

One major problem in making calculations is to know where the people will be at the moment the bombs explode. Calculations for the United States are generally based on the 1970 census, but it should be borne in mind that the census data describe where people's homes are, and there is never a moment when everybody in the United States is at home at the same time. If an attack took place during a working day, casualties might well be higher since people would be concentrated in factories and offices (which are more likely to be targets) rather than dispersed in the suburbs. For the case of the Soviet population, the same assumption is made that people are at home, but the inaccuracies are compounded by the unavailability of detailed information about just where the Soviet rural population lives. The various calculations that were used made varying, though not unreasonable, assumptions about population location.

115

Hiroshima cinema, which stood on a busy street 850 meters east of the hypocenter. It was crushed out of shape by blast and heat. (Shunkichi Kikuchi, Hiroshima–Nagasaki Publishing Committee.)

A second uncertainty in calculations has to do with the degree of protection available. There is no good answer to the question: "Would people use the best available shelter against blast and fallout?" It seems unreasonable to suppose that shelters would not be used and equally unreasonable to assume that at a moment of crisis all available resources would be put to rational use. (It has been pointed out that if plans worked, people behaved rationally, and machinery were adequately maintained, there would be no peacetime deaths from traffic accidents.) The Defense Civil Preparedness Agency has concluded from public opinion surveys that in a period of severe international crisis about 10 percent of all Americans would leave their homes and move to a "safer"

To illustrate the vulnerability of the economy in a nuclear war, two one-megaton warheads would destroy the Baytown refinery in Baytown, Texas, which supplies approximately 3.5 percent of the total U.S. oil-refining capacity. Eighty one-megaton warheads would destroy 64 percent of the total refining capacity for the country. Similarly, sixty-four 40-kiloton and nine 170-kiloton warheads would destroy 73 percent of the total USSR petroleum-refining capacity.

Data from the Office of Technology Assessment

place (spontaneous evacuation); more reliable estimates are probably impossible, but it could make a substantial difference to the casualty figures.

A third uncertainty is the weather at the time of the attack at the various places where bombs explode. The local wind conditions, and especially the amount of moisture in the air, may make an enormous difference in the number and spread of fires. Wind conditions over a wider area determine the extent and location of fallout contamination.

The time of year has a decisive effect on the damage that fallout does to agriculture—while an attack in January might be expected to do only indirect damage (destroying farm machinery or the fuel to run it), fallout when plants are young can kill them, and fallout just before harvest time would probably make it unsafe to get the harvest in. The time of year also has direct effects on population death—an attack in the dead of winter, which might not directly damage agriculture, may lead to greater deaths from fallout radiation (because of the difficulty of improvising fallout protection by moving frozen dirt) and from cold and exposure.

The question of how rapid and efficient economic recovery would be—or indeed whether a genuine recovery would be possible at all—raises questions that seem to be beyond calculation. It is possible to calculate direct economic damage by making assumptions about the size and exact location of bomb explosions and the hardness of economic assets; however, such calculations cannot address the issues of bottlenecks and of synergy. Bottlenecks would occur if a key product that was essential for many other manufacturing processes could no longer be produced, or

117

(for the case of a large attack) if an entire industrial sector were wiped out. In either case, the economic loss would greatly exceed the peacetime value of the factories that were actually destroyed. There does not appear to be any reliable way of calculating the likelihood or extent of bottlenecks because economic input/output models do not address the possibility or cost of substitutions across sectors.

Apart from the creation of bottlenecks, there could be synergistic effects: for example, the fire that cannot be controlled because the blast destroyed fire stations, as actually happened at Hiroshima. Here, too, there is no reliable way to estimate the likelihood of such effects: would radiation deaths of birds and the destruction of insecticide factories have a synergistic effect?

Another uncertainty is the possibility of organizational bottlenecks. In the most obvious instance, it would make an enormous difference whether the President of the United States survived. Housing, defined as a place where a productive worker lives as distinct from shelter for refugees, is another area of uncertainty. Minimal housing is essential if production is to be restored, and it takes time to rebuild it if the existing housing stock is destroyed or is beyond commuting range of the surviving (or repaired) workplaces. The United States has a much larger and more dispersed housing stock than does the Soviet Union, but American workers have higher minimum standards.

There is a final area of uncertainty that this study does not even address, but which could be of very great importance. Actual nuclear attacks, unlike those in this study, would not take place in a vacuum. There would be a series of events that would lead up to the attack, and these events could markedly change both the physical and the psychological vulnerability of a population to a nuclear attack. Even more critical would be the events after the attack. Assuming that the war ends promptly, the terms on which it ends could greatly affect both the economic condition and the state of mind of the population. The way in which other countries are affected could determine whether the outside world is a source of help or of further danger. The post-attack military situation (and nothing in this study addresses the effects of nuclear attacks on military power) could determine not only the attitude of other countries, but also whether limited surviving resources are put to military or to civilian use.

General Description of Effects

The energy of a nuclear explosion is released in a number of different ways:

- an explosive blast, which is qualitatively similar to the blast from ordinary chemical explosions, but which has somewhat different effects because it is typically so much larger
- direct nuclear radiation
- direct thermal radiation, most of which takes the form of visible light
- pulses of electrical and magnetic energy, called electromagnetic pulse (EMP)
- the creation of a variety of radioactive particles, which are thrown up into the air by the force of the blast, and are called radioactive fallout when they return to earth

The distribution of the bomb's energy among these effects depends on its size and on the details of its design, but a general description is possible.

Blast

Most damage to cities from large weapons comes from the explosive blast. The blast drives air away from the site of the explosion, producing sudden changes in air pressure (called static overpressure) that can crush objects, and high winds (called dynamic pressure) that can move them suddenly or knock them down. In general, large buildings are destroyed by the overpressure, while people and objects such as trees and utility poles are destroyed by the wind.

For example, consider the effects of a one-megaton airburst on things 4 miles (6 kilometers) away. The overpressure will be in excess of 5 pounds per square inch (psi), which will exert a force of more than 180 tons on the wall of a typical two-story house. At the same place, there would be a wind of 160 mph (255 km/h). While 5 psi is not enough to crush a man, a wind of 180 mph could create fatal collisions between people and nearby objects.

The magnitude of the blast effect (generally measured in pounds per square inch) diminishes with distance from the center of the

At the instant of the explosion of a 15-kiloton atomic bomb, a blinding flash lights up a wood-frame house. Operation Doorstep, Yucca Flats, Nevada, March 17, 1953. (U.S. Federal Emergency Management Agency.)

Almost instantaneously, the surface begins to catch fire, Operation Doorstep. (U.S. Federal Emergency Management Agency.)

120

Then, $2\frac{1}{3}$ seconds after the explosion, the house is demolished (5-psi peak overpressure), Operation Doorstep. (U.S. Federal Emergency Management Agency.)

explosion. It is related in a more complicated way to the height of the burst above ground level. For any given distance from the center of the explosion, there is an optimum burst height that will produce the greatest overpressure, and the greater the distance the greater the optimum burst height. As a result, a burst on the surface produces the greatest overpressure at very close ranges (which is why surface bursts are used to attack very hard, very small targets such as missile silos), but less overpressure than an airburst at somewhat longer ranges. Raising the height of the burst reduces the overpressure directly under the bomb, but widens the area at which a given smaller overpressure is produced. Thus, an attack on factories with a one-megaton weapon might use an airburst at an altitude of 8000 feet (2400 meters), which would maximize the area (about 28 square miles; 7200 hectares) that would receive 10 psi or more of overpressure.

When a nuclear weapon is detonated on or near the surface of the earth, the blast digs out a large crater. Some of the material that used to be in the crater is deposited on the rim of the crater;

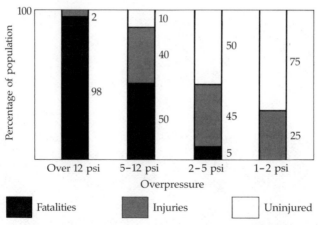

Vulnerability of population in various overpressure zones. (Office of Technology Assessment.)

the rest is carried up into the air and returns to the earth as fall-out. An explosion that is farther above the earth's surface than the radius of the fireball does not dig a crater and produces negligible immediate fallout.

For the most part, blast kills people by indirect means rather than by direct pressure. While a human body can withstand up to 30 psi of simple overpressure, the winds associated with as little as 2 to 3 psi could be expected to blow people out of typical modern office buildings. Most blast deaths result from the collapse of occupied buildings, from people being blown into objects, or from buildings or smaller objects being blown onto or into people. Clearly, then, it is impossible to calculate with any precision how many people would be killed by a given blast—the effects would vary from building to building.

In order to estimate the number of casualties from any given explosion, it is necessary to make assumptions about the proportion of people who will be killed or injured at any given overpressure. The assumptions used in this chapter are shown in the figure above. They are relatively conservative. For example, weapons tests suggest that a typical residence will be collapsed by an overpressure of about 5 psi. People standing in such a residence have a 50 percent chance of being killed by an overpressure of 3.5 psi, but people who are lying down at the moment the blast

A residential quarter on a hill 150–300 meters from the hypocenter in Nagasaki (20–30 psi peak overpressure). The houses collapsed from the blast, and the fire consumed them all. Not a soul survived in this area. (Shigeo Hayashi, Hiroshima–Nagasaki Publishing Committee.)

wave hits have a 50 percent chance of surviving a 7-psi overpressure. The calculations used here assume a mean lethal overpressure of 5 to 6 psi for people in residences, meaning that more than half of those whose houses are blown down on top of them will nevertheless survive. Some studies use a simpler technique: they assume that the number of people who survive in areas receiving more than 5 psi equals the number of people killed in areas receiving less than 5 psi, and hence that fatalities are equal to the number of people inside a 5-psi ring.

Unreinforced brick house, 4700 feet from ground zero before a nuclear explosion at the Nevada test site. Operation Cue. (U.S. Federal Emergency Management Agency.)

Unreinforced brick house after the nuclear explosion (5-psi peak overpressure), Operation Cue. (U.S. Federal Emergency Management Agency.)

Direct Nuclear Radiation

Nuclear weapons inflict ionizing radiation on people, animals, and plants in two different ways. Direct radiation occurs at the time of the explosion; it can be very intense, but its range is limited. Fallout radiation is received from particles that are made radioactive by the effects of the explosion and subsequently distributed at varying distances from the site of the blast. [Fallout is discussed in Chapter 18.]

For large nuclear weapons, the range of intense direct radiation is less than the range of lethal blast and thermal radiation effects. However, in the case of smaller weapons, direct radiation may be the lethal effect with the greatest range. Direct radiation did substantial damage to the residents of Hiroshima and Nagasaki.

Human response to ionizing radiation is subject to great scientific uncertainty and intense controversy. It seems likely that even small doses of radiation do some harm. To understand the effects of nuclear weapons, one must distinguish between short-term and long-term effects.

Short-Term Effects. A dose of 600 rems within a short period of time (six to seven days) has a 90 percent chance of creating a fatal illness, with death occurring within a few weeks. (A *rem* or "roentgen-equivalent-man" is a measure of biological damage; a *rad* is a measure of radiation energy absorbed; a *roentgen* is a measure of radiation energy; for our purposes it may be assumed that 100 roentgens is equivalent to 100 rads or 100 rems.) [Editors' note: Some maintain that 600 rems of whole-body radiation is a fatal dose to 90 percent of exposed adults only if it is given in a period of one day or less and that if the exposure is spread over six to seven days then the 90 percent lethal dose is significantly greater. At the present time, there is insufficient evidence to establish a clear relationship between the rate of radiation exposure and mortality.]

The precise shape of the curve showing the death rate as a function of radiation dose is not known in the region between 300 and 600 rems, but a dose of 450 rems within a short time is estimated to create a fatal illness in half the people exposed to it; the other half would get very sick, but would recover. A dose of 300 rems might kill about 10 percent of those exposed. A dose of

200 to 450 rems will cause a severe illness from which most people would recover; however, this illness would render people highly susceptible to other diseases or infections. [See Chapter 16.] A dose of 50 to 200 rems will cause nausea and lower resistance to other diseases, but medical treatment is not required. A dose below 50 rems will not cause any short-term effects that the victim will notice, but will nevertheless do long-term damage. [Editors' note: These doses are for whole-body radiation, and the percentage mortality figures refer to a healthy adult population.]

Long-Term Effects. The effects of smaller doses of radiation are long term and measured in a statistical way. A dose of 50 rems generally produces no short-term effects; however, if a large population were exposed to 50 rems, somewhere between 0.4 and 2.5 percent of them would be expected to contract fatal cancer (after some years) as a result. There would also be serious genetic effects for some fraction of those exposed. Lower doses produce lower effects. There is a scientific controversy about whether any dose of radiation, however small, is really safe. [Chapter 18 discusses the extent of the long-term effects that a nuclear attack might produce.] It should be clearly understood, however, that a large nuclear war would expose the survivors, however well sheltered, to levels of radiation far greater than the U.S. government considers safe in peacetime.

Thermal Radiation

Approximately 35 percent of the energy from a nuclear explosion is an intense burst of thermal radiation, i.e., heat. The effects are roughly analogous to the effect of a 2-second flash from an enormous sunlamp. Since the thermal radiation travels at the speed of light (actually a bit slower, since it is deflected by particles in the atmosphere), the flash of light and heat precedes the blast wave by several seconds, just as lightning is seen before the thunder is heard.

The visible light will produce "flashblindness" in people who are looking in the direction of the explosion. Flashblindness can last for several minutes, after which recovery is total. A one-megaton explosion could cause flashblindness at distances as great as 13 miles (21 kilometers) on a clear day or 53 miles (85

Flash shadows at Nagasaki. A soldier descended from an observation post as the "all clear" was sounded. He unfastened his sword, hooked it onto a clapboard, and unbuttoned his jacket when he saw the flash. His shadow is burned into the plank wall (3.5 kilometers from the hypocenter). (Eiichi Matsumoto, Hiroshima–Nagasaki Publishing Committee.)

kilometers) on a clear night. If the flash is focused through the lens of the eye, a permanent retinal burn will result. At Hiroshima and Nagasaki, there were many cases of flashblindness, but only one case of retinal burn, among the survivors. On the other hand, anyone flashblinded while driving a car could easily cause permanent injury to himself and to others.

Skin burns result from higher intensities of light, and therefore take place closer to the point of explosion. A one-megaton explosion can cause first-degree burns (equivalent to a bad sunburn) at distances of about 7 miles (11 kilometers), second-degree burns (producing blisters that lead to infection if untreated, and permanent scars) at distances of about 6 miles (10 kilometers), and third-degree burns (which destroy skin tissue) at distances of up to 5 miles (8 kilometers). Third-degree burns over 24 percent of the body, or second-degree burns over 30 percent of the body, will result in serious shock, and will probably prove fatal unless prompt, specialized medical care is available. The entire United States has facilities to treat 1000 or 2000 severe burn cases; a single nuclear weapon could produce more than 10,000.

The distance at which burns are dangerous depends heavily on weather conditions. Extensive moisture or a high concentration of particles in the air (smog) absorbs thermal radiation. Thermal radiation behaves likes sunlight, so objects create shadows behind which the thermal radiation is indirect (reflected) and less intense. Some conditions, such as ice on the ground or low white clouds over clean air, can increase the range of dangerous thermal radiation.

Fires

The thermal radiation from a nuclear explosion can directly ignite kindling materials. In general, ignitible materials outside the house, such as leaves or newspapers, are not surrounded by enough combustible material to generate a self-sustaining fire. Fires more likely to spread are those caused by thermal radiation passing through windows to ignite beds and overstuffed furniture inside houses. A rather substantial amount of combustible material must burn vigorously for 10 to 20 minutes before the room or whole house becomes inflamed. The blast wave, which arrives after most thermal energy has been expended, will have some extinguishing effect on the fires. However, studies and tests

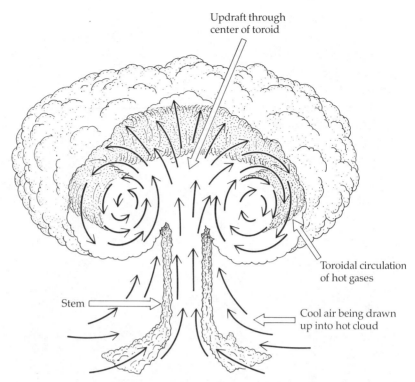

Updraft through
center of toroid

Toroidal circulation
of hot gases

Stem

Cool air being drawn
up into hot cloud

Air circulation during mushroom cloud formation after a nuclear explosion. The negative pressure caused by the updraft creates hurricane winds moving back toward the hypocenter, which would add to the possibility of firestorm. (From S. Glasstone and P. J. Dolan, 1977, The Effects of Nuclear War, *3rd ed., U.S. Department of Defense and U.S. Department of Energy.)*

of this effect have been very contradictory, so the extent to which blast can be counted on to extinguish fire starts remains quite uncertain.

Another possible source of fires, which might be more damaging in urban areas, is indirect. Blast damage to stoves, water heaters, furnaces, electrical circuits, or gas lines would ignite fires where fuel is plentiful.

The best estimates are that at the 5-psi level about 10 percent of all buildings would sustain a serious fire, while at 2 psi about 2 percent would have serious fires, usually arising from secondary sources such as blast-damaged utilities rather than direct thermal radiation.

Dresden after the firestorm. More than 35,000 people died. (Ullstein Bilderdienst, Berlin.)

"*Dresden burned for a week. The worst fire came within an hour after the British attack, when thousands of separate blazes merged into a howling fire storm of the type created by the RAF at Hamburg in July of 1943. It engulfed some 1600 acres, practically the whole of old Dresden, generating winds of tornado force, incinerating everything and everyone in its path, and sucking the life out of those who had attempted to seek refuge in the cellars of the city.*" (From **The Air War in Europe**, World War II, *Time-Life Books, 1979.*)

It is possible that individual fires, whether caused by thermal radiation or by blast damage to utilities, furnaces, etc., would coalesce into a mass fire that would consume all structures over a large area. This possibility has been intensely studied, but there remains no basis for estimating its probability. Mass fires could be of two kinds: a "firestorm," in which violent, inrushing winds create extremely high temperatures but prevent the fire from spreading radially outward, and a "conflagration," in which a fire spreads along a front.

Hamburg, Tokyo, and Hiroshima experienced firestorms in World War II; the Great Chicago Fire and the San Francisco Earthquake Fire were conflagrations. A firestorm is likely to kill a high proportion of the people in the area of the fire, through heat and through asphyxiation of those in shelters. A conflagration spreads slowly enough so that people in its path can escape, though a conflagration caused by a nuclear attack might take a heavy toll of those too injured to walk.

Some believe that firestorms in U.S. or Soviet cities are unlikely because the density of flammable materials ("fuel loading") is too low—the ignition of a firestorm is thought to require a fuel loading of at least 8 lb/ft^2 (Hamburg had 32), compared to fuel loading of 2 lb/ft^2 in a typical U.S. suburb and 5 lb/ft^2 in a neighborhood of two-story brick rowhouses. The likelihood of a conflagration depends on the geography of the area, the speed and direction of the wind, and details of building construction. Another variable is whether people and equipment are available to fight fires before they can coalesce and spread. [Editors' note: The presence of large stocks of combustible fuel in U.S. cities may increase the likelihood of firestorms dramatically. Even for large conflagrations, however, the problems of high surface temperatures and oxygen depletion would make most shelter survival impossible.]

Electromagnetic Pulse

Electromagnetic pulse (EMP) is an electromagnetic wave similar to radio waves, which results from secondary reaction occurring when the nuclear gamma radiation is absorbed in the air or ground. It differs from the usual radio waves in two important ways. First, it creates much higher electric field strengths than the thousands of volts a radio signal might produce. Second, it is a single pulse of energy that disappears completely in a small fraction of a second. In this sense, it is rather similar to the electrical signal from lightning, but the rise in voltage is typically a hundred times faster. This means that most equipment designed to protect electrical facilities from lightning works too slowly to be effective against EMP.

The strength of an EMP is measured in volts per meter and is an indication of the voltage that would be produced in an exposed antenna. A nuclear weapon burst on the surface will typically produce an EMP of tens of thousands of volts per meter at short distances (10-psi range) and thousands of volts per meter at longer distances (1-psi range). Airbursts produce less EMP, but high-altitude bursts (above 19 miles (30 kilometers)) produce very strong EMP, with ranges of hundreds or thousands of miles. An attacker might detonate a few weapons at such altitudes in an effort to destroy or damage the communications and electric power systems of the victim.

There is no evidence that EMP is a physical threat to humans. However, electrical or electronic systems, particularly those connected to long wires such as powerlines or antennas, can undergo either of two kinds of damage. First, there can be actual physical damage to an electrical component such as shorting of a capacitor or burnout of a transistor, which would require replacement or repair before the equipment can again be used. Second, at a lesser level, there can be a temporary operational upset, frequently requiring some effort to restore operation. For example, instabilities induced in power grids can cause the entire system to shut itself down, upsetting computers that must be started again. Base radio stations are vulnerable not only from the loss of commercial power, but from direct damage to electronic components connected to the antenna. In general, portable radio transmitter/ receivers with relatively short antennas are not susceptible to EMP. The vulnerability of the telephone system to EMP could not be determined.

Fallout

While any nuclear explosion in the atmosphere produces some fallout, the fallout is far greater if the burst is on the surface, or at least low enough for the fireball to touch the ground. The fallout from airbursts alone poses long-term health hazards, but they are trivial compared to the other consequences of a nuclear attack.

132

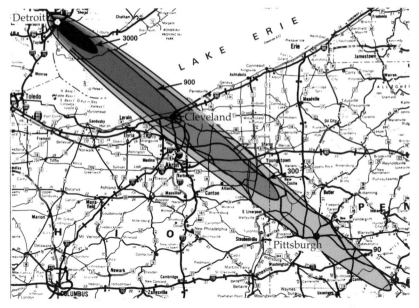

Main fallout pattern from a one-megaton surface burst in Detroit, assuming a uniform 15-mph northwest wind. The contours represent 7-day accumulated doses (without shielding) of 3000, 900, 300, and 90 rems. (Office of Technology Assessment.)

The significant hazards come from particles scooped up from the ground and irradiated by the nuclear explosion.

The radioactive particles that rise only a short distance (those in the "stem" of the familiar mushroom cloud) will fall back to earth within a matter of minutes, landing close to the center of the explosion. Such particles are unlikely to cause many deaths, because they will fall in areas where most people have already been killed. However, the radioactivity will complicate efforts at rescue or eventual reconstruction.

The radioactive particles that rise higher will be carried some distance by the wind before returning to the earth, and hence the area and intensity of the fallout are strongly influenced by local weather conditions. Most of the material is simply blown downwind in a long plume. The map shown in the figure above illustrates the plume expected from a one-megaton surface burst in Detroit, if winds were blowing from the northwest. The illustrated plume assumed that the winds were blowing at a uniform speed of 15 mph (24 km/h) over the entire region. The plume would be longer and thinner if the winds were more intense, and shorter and somewhat more broad if the winds were slower. Enough fallout would deposit on Cleveland to inflict acute radiation sickness on those who did not evacuate or use effective fallout shelters. [Editors' note: Some sections of Cleveland under the conditions given would have seven-day accumulated dose exposures of 900 rems, a fatal dose for all those in the open.]

Wind direction can make an enormous difference. Rainfall can also have a significant influence on the ways in which radiation from smaller weapons is deposited, since rain will carry contaminated particles to the ground. The areas receiving such contaminated rainfall would become "hot spots," with greater radiation intensity than their surroundings. When the radiation intensity from fallout is great enough to pose an immediate threat to health, fallout will generally be visible as a thin layer of dust.

The amount of radiation produced by fallout materials will decrease with time as the radioactive materials "decay." Each material decays at a different rate. Materials that decay rapidly give off intense radiation for a short period of time, while long-lived materials radiate less intensely, but for longer periods. Immediately after the fallout is deposited in regions surrounding the blast site, radiation intensities will be very high as the short-lived materials decay. These intense radiations will decrease relatively quickly. The areas in the plume illustrated in the figure would become "safe" (by peacetime standards) in two to three years for the outer ellipse, and in ten years or so for the inner ellipse.

Some radioactive particles will be thrust into the stratosphere and may not return to earth for some years. In this case, only the particularly long-lived particles pose a threat, and they are dispersed around the world over a range of latitudes. Some fallout from U.S. and Soviet weapons tests in the 1950's and early 1960's can still be detected. There are also some particles in the immediate fallout (notably strontium-90 and cesium-137) that remain radioactive for years.

The biological effects of fallout radiation are substantially the same as those from direct radiation, discussed earlier. People exposed to enough fallout radiation will die, and those exposed to lesser amounts may become ill.

There is some public interest in the question of the consequences if a nuclear weapon destroyed a nuclear power plant. The core of a power reactor contains large quantities of radioactive material, which tends to decay more slowly (and hence less intensely) than the fallout particles from a nuclear weapons explosion. Consequently, fallout from a destroyed nuclear reactor (whose destruction would, incidentally, require a high-accuracy surface burst) would not be much more intense (during the first day) or widespread than "ordinary" fallout, but would stay radioactive for a considerably longer time. Areas receiving such fallout would have to be evacuated or decontaminated; otherwise, survivors would have to stay in shelters for months. [Editors' note: See the figure on p. 39, Chapter 4.]

Combined Injuries (Synergism)

So far, the discussion of each major effect (blast, nuclear radiation, and thermal radiation) has explained how this effect in isolation causes deaths and injuries to humans. It is customary to calculate the casualties accompanying hypothetical nuclear explosion as follows: for any given range, the effect most likely to kill people is selected and its consequences calculated, while the other effects are ignored. It is obvious that combined injuries are possible, but there are no generally accepted ways of calculating their probability. What data do exist seem to suggest that calculations of single effect are not too inaccurate for immediate deaths, but that deaths occurring some time after the explosion may well be due to combined causes, and hence are omitted from most calculations.

Nuclear Radiation Combined with Thermal Radiation. Severe burns place considerable stress on the blood system and often cause anemia. It is clear from experiments with laboratory animals that exposure of a burn victim to more than 100 rems of radiation will impair the blood's ability to support recovery from the thermal burns. Hence a sublethal radiation dose could make it impossible to recover from a burn that, without the radiation, would not cause death.

Nuclear Radiation Combined with Mechanical Injuries. Mechanical injuries, the indirect results of blast, take many forms. Flying glass and wood will cause puncture wounds. Winds may blow people into obstructions, causing broken bones, concussions, and

135

internal injuries. Persons caught in a collapsing building can suffer many similar mechanical injuries. There is evidence that all of these types of injuries are more serious if the person has been exposed to 300 rems, particularly if treatment is delayed. Blood damage will clearly make a victim more susceptible to blood loss and infection. This has been confirmed in laboratory animals in which a borderline lethal radiation dose was followed a week later by a blast overpressure that alone would have produced a low level of prompt lethality. The number of prompt and delayed (from radiation) deaths both increased over what would be expected from the single effect alone.

Thermal Radiation and Mechanical Injuries. There is no information available about the effects of this combination, beyond the common sense observation that since each can place a great stress on a healthy body, the combination of injuries that are individually tolerable may subject the body to a total stress that it cannot tolerate. Mechanical injuries should be prevalent at about the distance from a nuclear explosion that produces sublethal burns, so this synergism could be an important one.

In general, synergistic effects are most likely to produce death when each of the injuries alone is quite severe. Because the uncertainties of nuclear effects are compounded when one tries to estimate the likelihood of two or more serious but (individually) nonfatal injuries, there really is no way to estimate the number of victims.

A further dimension of the problem is the possible synergy between injuries and environmental damage. To take one obvious example, poor sanitation (due to the loss of electrical power and water pressure) can clearly compound the effects of any kind of serious injury. Another possibility is that an injury would so immobilize the victim that he would be unable to escape from a fire.

10

The Medical Effects on a City in the United States

H. Jack Geiger, M.D.

The prospect of apocalypse strips words of their ordinary meanings. We recognize that the detonation of only a single medium-sized thermonuclear weapon over a major metropolitan area in the United States (or any other nation) would produce death, injury, destruction, and devastation on a scale that is without precedent in the history of mankind. Yet to speak of the effects of such an assault on human beings as "medical," while it may be a necessary distortion, is to demonstrate the insufficiency of language and the inadequacy of all past human experience to the realities of nuclear conflict.

Those realities create a curious dichotomy. On the one hand, we can describe the consequences of a thermonuclear explosion in physical and environmental terms and even, within broad limits, express them scientifically: blast pressures in pounds per square inch, heat in calories per square centimeter, radiation in units of type and intensity, all resulting in specific numbers of deaths and profound injuries. On the other hand, we can neither fully imagine nor truly comprehend the event we are describing, for it defies comparison—qualitatively and quantitatively—with anything we have already known.

The difficulty of comprehension is particularly applicable to the United States, a nation that has been spared the direct experi-

> I would say when we start to talk about the megatonnage we could bring into a nuclear war, we are talking about annihilation. How many times do you have to hit a target with nuclear weapons?
>
> John F. Kennedy
> December 17, 1962

ence even of conventional war for more than a century. Occasional natural or man-made disasters—the Great Chicago Fire, the San Francisco Earthquake, the Texas City ship explosion—have been of limited duration and effect, well within the capacity of an intact and thriving society to repair with help from outside. Although agencies of the U.S. government have, for three decades, prepared detailed scenarios of nuclear attack—even calculating the probable casualties for every American community of 25,000 or more inhabitants—these have not until recently been widely publicized or discussed. The resultant irony is that the only nation that has ever used nuclear weapons on a civilian population is the one that is least prepared, by direct past experience of war or disaster, to comprehend their effects. On the other hand, Americans are among the best informed about the physical effects of nuclear weapons and the medical consequences of their use as a result of the massive educational effort on the part of the scientific and medical community as well as certain government agencies.

The most probable scenarios of so-called countervalue attacks targeted on U.S. population centers, or even of more limited "counterforce" attacks on missile bases, military installations, and industrial centers, call for massive, multi-weapon strikes. In such attacks, cities like New York would suffer 20 to 35 nuclear warhead explosions. Other metropolitan centers would be hit at a rate of one thermonuclear weapon for every 500,000 or even 200,000 residents, with totals of more than 6000 megatons nationwide. The resulting levels of destruction would be so great as to be almost incalculable.

To make the effects of nuclear war more comprehensible, therefore, military planners and civil defense specialists usually examine the effects of a single weapon on a single target. It is a measure of the terrifying growth of nuclear arsenals over the last 30 years that the standard unit of nuclear currency for such calculations is a one-megaton weapon, only a medium-sized warhead in present-day stockpiles but 80 times more powerful than the weapon that leveled Hiroshima.

To estimate the "medical" consequences of such a one-megaton nuclear explosion, we must begin by specifying the variables that influence the location, range, and type of a weapon's physical effects, for these obviously are major determinants of the fate of people in the target area. At a given megatonnage, an airburst distributes blast and heat effects over a wider area; a groundburst will concentrate blast and heat more intensely at ground zero, shorten the range of blast and heat effects, but add the hazard of prompt, intense, and often lethal local radioactive fallout. Local atmospheric conditions make a major difference: wind velocity and direction determine the distribution of fallout and influence the number and spread of fires. Increased moisture in the atmosphere will diminish fire damage, but active precipitation will greatly increase fallout; fog, smog, and clouds will reduce visibility, absorb heat, and decrease the number of burns.

Other variables are directly related to human social behavior. An attack during the summer is likely to find more people outdoors, lightly clothed, and therefore more vulnerable to the effects of blast and heat. An attack on a weekday, and especially during working hours, will find additional tens or hundreds of thousands of people (or, in very large cities, millions) concentrated in offices, factories, and schools in central city areas. The most devastating one-megaton explosion, therefore, might be an airburst at 6000 feet (the altitude that maximizes the area receiving at least 20 pounds per square inch of blast-induced overpressure, and therefore maximizes death and injury even in the strongest buildings), occurring during weekday working hours on a clear, dry, windy day in summer, with ground zero targeted in the heart of the central-city business district.

Finally, the amount of destruction—and the number of deaths and injuries—will depend in part on variables that cannot be predicted accurately. If a mass conflagration develops—a large, moving fire fanned by surface winds—burn injuries and deaths will increase substantially. If a firestorm develops—a huge, intense but stationary fire, burning at temperatures of over 800°C, sucking in cooler air and creating winds of 200 miles per hour or

139

City population census figures do not include the additional hundreds of thousands or millions of people who come to cities for work, shopping, or recreation. New York City, September 9, 1981. (United Press International.)

140

more—the lethal area will be enlarged fivefold, and burn deaths will increase enormously.

The physical effects of such an explosion are usually expressed as a series of concentric circles around ground zero, each demarcating a zone of destruction in which the magnitudes of radiation, blast, and heat, and their effects on buildings, other structures, and people, can be roughly estimated. This movable nuclear bull's eye can also be used to express the "medical" effects. It applies equally to all cities, with differences due only to variations in population concentration and in the local terrain.

What if ground zero were the heart of downtown Detroit? What would happen in each of the circles?

The first circle, with a radius of 1.5 miles from ground zero, encompassing an area of 7 square miles, is the region of nearly total destruction—and total lethality. Blast overpressures ranging from 200 pounds per square inch (psi) down to 20 psi would crush, collapse, or explode all buildings, including the most strongly constructed steel and reinforced-concrete structures. Winds of 600 miles per hour would hurl debris outward at lethal velocities. Temperatures in the fireball above ground zero would exceed 27 million degrees Fahrenheit, and in this area everything would be vaporized; elsewhere in the circle, the heat would melt glass and steel, and concrete would explode. Direct radiation would range from 11,000 rads near ground zero to 1100 at the circle's rim. All of the human beings in this circle would die almost immediately—vaporized, crushed, charred, or radiated. With no survivors, there would be no "medical" problems.

The second circle extends 2.9 miles from ground zero and encompasses an additional 20 square miles of Detroit. Blast overpressures are 10 to 20 psi, sufficient to collapse and destroy all but the strongest buildings and to sweep out the floors and walls of steel skyscrapers. Winds are 250 to 300 miles an hour, enough to hurl 180-pound adults 300 feet or more at high speeds. In this circle, 50 percent of the population would die of blast injuries

---

rad   unit of radiation; 450 rads in a short period of time is a lethal dose for approximately 50 percent of healthy adults exposed.

Detroit skyline.

alone: crushed chests and extremities, skull fractures, penetrating wounds of the chest and abdomen, ruptured lungs and other internal organs, crushed vertebrae and transected spinal cords, multiple lacerations and profound hemorrhage. All exposed persons who escaped death from blast would suffer third-degree burns unless they were temporarily shielded from the fireball by buildings that had not yet collapsed (the thermal pulse precedes the blast wave). The heat would evaporate aluminum siding, melt acrylic windows, and cause spontaneous ignition of clothing. In general, however, blast injuries would predominate; not enough buildings would remain to permit the development or spread of secondary fires.

The perimeter of the third circle, with a radius of 4.3 miles, marks the border of the lethal zone, the area in which the total population count is numerically the same as the total of fatalities

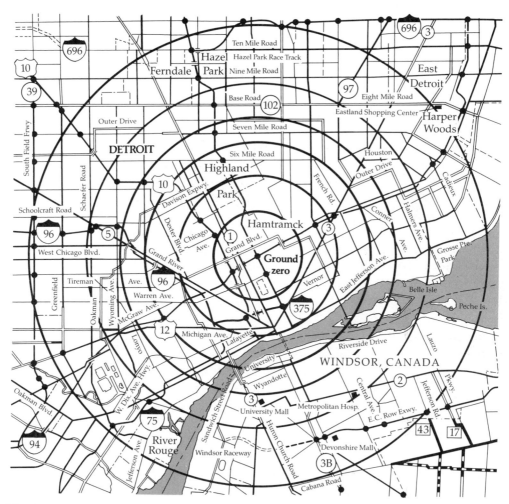

The six circles of a one-megaton airburst over Detroit. For a 20-megaton explosion, the sixth circle would be the first circle.

in all of Detroit. In this circle, which encompasses an additional 32 square miles, overpressures are 5 psi—enough to exert 180 tons of pressure on the wall of a two-story house—and winds are 160 miles per hour, capable of hurling adults 20 feet at 14 miles per hour. Factories and commercial buildings of heavy construction would be severely damaged. Debris—pieces of concrete, steel, rock, glass, and the like—would be traveling at lethal speeds. The heat—approximately 40 calories per square centimeter—would melt asphalt paving, ignite wood and fabrics (bedding, carpeting, curtains, upholstery) inside buildings, ignite clothing, and cause major fires in at least 10 percent of the buildings. Blast-induced trauma would still predominate, though all unprotected persons outdoors would suffer third-degree burns and many others would suffer flame burns. Most persons who made a reflex glance at the fireball would be flashblinded temporarily; many would suffer retinal burns and partial or total blindness; many, in addition, would be made deaf by rupture of the eardrums.

In these second and third rings, encompassing more than 50 square miles of Detroit's most densely populated areas, 50 percent of the population would die promptly and 40 percent would suffer serious and incapacitating injuries. Infected burns and untreated trauma, killing people slowly, would add many of the injured to the total of fatalities 30 or more days later.

The fourth circle has a radius of 4.9 miles and adds 18 square miles to the area of damage. Overpressures of 4 psi and winds of 125 mph—greater than hurricane force—are sufficient to destroy brick and frame houses completely but would leave many stronger buildings standing, available to fuel the fires created not only by the thermal pulse (25 calories per square centimeter, still sufficient to ignite bedding and clothing and cause third-degree burns to unprotected skin) but also by broken furnaces, ruptured natural gas lines, downed high-voltage lines, exploding gasoline stations and fuel tanks, and dry trees and vegetation.

At the outer rim of the fifth circle, 6.3 miles from ground zero, 100-mph winds and 3-psi overpressures would still be sufficient to blow people out of buildings, sweep away the outer walls of skyscrapers, and cause all the varieties of trauma, but burns would predominate. The heat would be sufficient to give third-degree burns to four of every five unprotected persons outdoors, to ignite dry grass, newspapers, and leaves, and to melt neoprene-treated nylon rainwear. The rings defined by circles four and five together—70 square miles of Detroit—the fire hazard would be intense. Only 5 percent of the population would be killed promptly, but 45 percent would be seriously injured.

The sixth circle, finally, has a radius of 8.5 miles; it adds another 100 square miles to the area of damage, bringing the total to 227 square miles. Overpressures are 2 psi and winds 70 to 80 mph. Smaller pieces of debris would still be lethal missiles, and windows would fragment into glass shards traveling at speeds close to 100 mph; 30 percent of all trees and utility poles would be

144

The blast destroyed the automobile and streetcar in front of the Nagarekawa Methodist Church, 1 kilometer from the hypocenter. Hiroshima, late September 1945. In a nuclear war the transportation network would be destroyed in targeted areas, making access to the injured all but impossible. (Eiichi Matsumoto, Hiroshima–Nagasaki Publishing Committee.)

downed, blocking the roads; brick and frame residences would be moderately damaged. At 5 to 7 calories per square centimeter, every fifth person outdoors would suffer a third-degree burn—but 70 percent would suffer extensive second-degree burns.

The sixth circle is important for another reason. If this were a 20-megaton explosion, this circle would be the first circle—a 227-square-mile area of total destruction and 100 percent fatalities.

Under the hypothesized conditions, and assuming that a firestorm developed, this single one-megaton explosion would kill 939,000 people—or 24 percent of the 3.86 million people in the Detroit metropolitan area. Another 1,145,000 would be seriously injured—almost every third person. Total casualties would be slightly more than two million, or 54 percent of the total. But even these are not the true totals; to them must be added substantial numbers of people who would become acutely ill or die as the supplies of insulin or digitalis or cortisone on which their lives depended became unavailable. Patients who required renal dialysis would die of uremia; cardiac pacemakers would be destroyed by the bomb's electromagnetic pulse. And the normal incidence of myocardial infarctions, perforated gastric ulcers, and other life-threatening emergencies would continue to occur among the uninjured.

But even the conservative estimates are beyond anything in all recorded human experience. Never have more than one million people with profound and incapacitating injuries been located in one place at one time—injuries, furthermore, that require the most complex and technologically sophisticated medical care for effective treatment. The number of victims with third-degree burns alone is likely to exceed 200,000—or 100 times more than all the intensive-care burn beds in the United States. The management of these and other injuries would require x-ray units, CT scanners, sterile burn treatment units, modern operating rooms, intensive-care units with sophisticated life support systems, and literally millions of units of blood and plasma, huge and varied supplies of drugs, and other specialized resources.

Similarly, the treatment of these 1.1 million wounded would require the efforts of tens of thousands of physicians, nurses, skilled technicians, and other support personnel.

But the thermonuclear explosion that has—in a few hours or days—created this incredible need for medical care has simultaneously obliterated the medical care system. Who would be left to respond—and with what resources?

In Detroit, as in most U.S. cities, physicians' offices are clustered in downtown areas closest to the putative ground zero. If anything, therefore, deaths and injuries to physicians and other health workers would occur at rates *greater* than those for the general population. Estimates of the ratios of surviving physicians to seriously injured victims vary from 1:350 to 1:1700. If we assume a ratio of 1:1000, and imagine that every surviving physician would find all of the wounded with no loss of time, spend only 15 minutes per patient on every aspect of diagnosis and treatment, and work 18 hours a day, it would still be 8 to 16 days before every surviving patient would be seen for the first time. Most of the victims, obviously, would die—many of them slowly and agonizingly—without any medical care, without even narcotics for the relief of their pain.

In reality, these are absurdly optimistic assumptions. Many physicians and patients would never find each other because of their fear of radiation exposure, because streets filled with rubble would make travel impossible, because victims would be trapped deep within wrecked buildings. The distribution would be un-

146

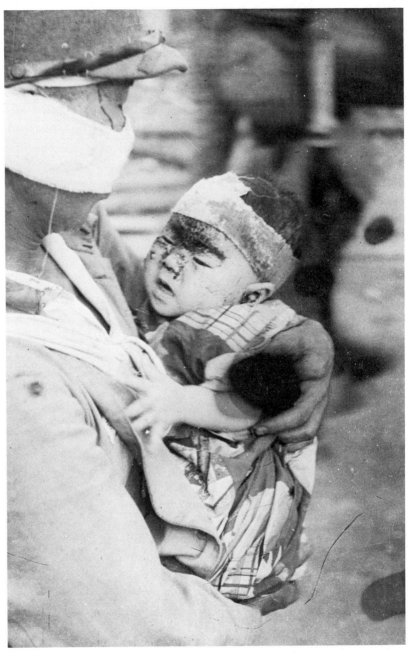

Nagasaki, about 2 PM on August 10, 1945, near the railway station. The baby, probably now motherless, is held firmly in the arms of its wounded father. Pain and tension on the infant's face express its anguish. (Yosuke Yamahata, Hiroshima–Nagasaki Publishing Committee.)

even: some physicians would face thousands of injured while many others would find none. There will be no communications system, no transportation network, no electricity, no water supply. Ambulances and other emergency vehicles would be nonexistent. In some areas, rubble and debris would rise to heights of 10 feet or more. Physicians and other medical personnel would be wracked by conflicting demands—the need to protect themselves and look after their own families and the responsibility to meet the overwhelming human demand for relief of pain and suffering.

But in "medical" terms, these ratios and calculations are meaningless as well, for the ministrations of a single physician or health worker, equipped only with his or her black bag and without diagnostic or therapeutic resources and facilities, can have no effect on injuries of this severity. Triage—the sorting out of those who might be saved from those who are doomed—would in most cases be impossible, and health workers would face an endless series of agonizing ethical decisions: for example, who should have priorities for treatment or access to miniscule supplies of drugs.

And what of the treatment resources? Hospitals—like physicians—are clustered in the zones of greatest probable damage and destruction. In the Detroit metropolitan area with 63 hospitals totaling approximately 18,000 beds, the entire medical care system would be overwhelmed by 1.1 million severely injured victims even if all personnel and support systems were intact and the physical environment was untouched. But in fact, only 5000 beds would remain undamaged—all outside the regions of greatest damage and farthest from the greatest concentrations of the injured. Conservatively, 80 percent of these beds would already be occupied, leaving a grand total of 1000 available beds—for each of which there would be 1000 profoundly injured patients.

A series of recent case studies makes it clear that the Detroit case is typical. Consider, for example, these data on the consequences of a one-megaton airburst over three other U.S. cities:

Philadelphia Metropolitan Area: Total population: 4,557,000. Killed: 769,000, or 17 percent of the total. Seriously injured: 1,334,000, or 29 percent. Total hospital beds in Philadelphia and nearby Camden, N.J.: 12,417, of which 1046 beds (8 percent) might remain functional. Total beds in six surrounding counties: 7516, yielding

148

8562 intact beds overall. Assuming 80 percent pre-attack occupancy, total available beds: 1730. Ratio of wounded to beds: 771 to 1.

Seattle Metropolitan Area: Total population: 1,213,000. Killed: 398,400, or 33 percent of the total. Seriously injured: 326,400, or 27 percent. Total hospital beds in the metropolitan area: 9483, of which only 1746 (18 percent) will be intact. Assuming 80 percent pre-attack occupancy, total available beds: 349. Ratio of wounded to beds: 935 to 1.

Lancaster, Pa., Area: Total population: 362,000. Killed: 150,000, or 41 percent of the total. Seriously injured: 35,000, or 9.7 percent. Total beds in the city and county: 1334, of which only 226 will remain intact. Assuming 80 percent pre-attack occupancy, total available beds: 45. Ratio of wounded to beds: 777 to 1.

The training of all physicians, and the daily experience of many of them, involves confrontation with human suffering and death—but suffering and death on this scale are qualitatively different, and it is no longer reasonable to think of the problem as "medical." In every U.S. metropolitan area, there are thousands of physicians with specialized training, extraordinary competence, and long experience in the management of severe trauma and severe burns, and they have at their disposal the nation's most complex and sophisticated hospital facilities. Every major American city, furthermore, has a medical disaster plan under which every physician, every health worker, and all the resources of hospitals, transportation and communications networks, police, fire departments, and other support systems can rapidly be mobilized into a single efficient team.

But—as our calculations for these four U.S. cities indicate—none of this would matter. The medical care would be overwhelmed by the consequences of a single thermonuclear weapon, let alone a substantial nuclear exchange. There can be no adequate medical response to a thermonuclear attack.

Much more is revealed by this analysis, however, than the collapse of medical care. Beyond the damage to human bodies, beyond the damage to the physical environment, what is illustrated is the destruction of a social system, a part of the intricate social fabric upon which a modern industrial society depends. In the wake of a thermonuclear attack, even simple activities—finding food and water, caring for children, obtaining shelter, preserving family groups—would be difficult or impossible. For individuals, biological survival is possible; for human populations, survival is a social as well as a biological phenomenon, and the ultimate wound is the rupture of the social fabric. For that wound, "medical" care—even if it were available—would be irrelevant.

References

Ervin, F. R., J. B. Glazier, S. Aronow, D. Nathan, R. Coleman, N. C. Avery, S. Shohet, and C. Leeman. 1962. "Human and Ecologic Effects in Massachusetts of an Assumed Thermonuclear Attack on the United States." In Symposium on the Medical Consequences of Thermonuclear War. *New England Journal of Medicine*, 266:1126–1155.

Glasstone, S., and P. J. Dolan. 1977. *The Effects of Nuclear Weapons*, 3rd ed. Washington, D.C.: U.S. Department of Defense and U.S. Department of Energy.

Office of Technology Assessment. 1979. *The Effects of Nuclear War*. Washington, D.C.: U.S. Congress.

11

Evaluation of the Medical Consequences of a Nuclear Attack

Leonid A. Ilyin, M.D.

The principal danger during the present stage of the nuclear arms race has been the enormous technical innovation in weapons systems, which has made a qualitative as well as a quantitative alteration in the nature of nuclear war. It is of crucial importance that the effects of a nuclear war be understood so that dangerous myths about the possibility of a "limited" nuclear war, and about surviving and winning a nuclear war, are dispelled.

The scale and nature of the consequences of nuclear war for a population depend on a number of diverse factors and circumstances. These can be divided into three categories, those that are predictable and quantifiable, those that are difficult to predict, and those that are unpredictable. Although it is possible to define many of the effects of a future nuclear war, at least half of the potential consequences are still not fully understood.

Those immediate consequences that can be described—mass death and incapacitation from the physical aspects of a nuclear explosion; the destruction of buildings, provisions, and fuel; the disruption of energy systems, water supplies, and sewage services—offer only a partial analysis of the whole tragedy of nuclear war for modern society.

Many of the predictable consequences can be estimated by using a mathematical model. Others, such as the interaction of the

World War II, Soviet woman returning home. In a nuclear war, persistent radiation would prohibit safe return to targeted areas for months to years. (Sovfoto.)

American soldiers, Buna, New Guinea, 1943. Even the massive number of dead and injured in World War II would seem insignificant compared to the casualties that would follow a nuclear war. (George Strock, Life, *from United Press International.)*

Stalingrad, February 1943. The inhabitants return to the burned-out city. Even these shells of buildings would not be standing in cities attacked by nuclear weapons; the destruction would be nearly total. (Sovfoto.)

various incapacitated elements of a modern society in the wake of a nuclear attack, are not possible to estimate. These difficult-to-predict and unpredictable effects make any calculations of the outcome of a nuclear war significant underestimates. In general, losses in population would depend on these factors (a whole range of other factors is not included):

- power and type of nuclear weapons
- method of explosion (air or surface burst)
- nature of the target (city or rural area) and the terrain
- density of population and its distribution in relation to the hypocenter
- degree of protection available to those exposed to the attack
- time of year and day, and meteorological conditions

London, December 29, 1940. The city suffered much damage during the Blitz, but a direct nuclear explosion in London would reduce practically all the buildings to rubble. (United Press International.)

It is necessary to differentiate between the direct and indirect effects of these weapons. The direct effects depend on direct exposure to the destructive energies of a nuclear explosion—the shock wave, thermal radiation, instantaneous radiation, and local radioactive fallout (if it is a surface explosion). These factors would produce mass death by various traumatic injuries (wounds and crushing syndromes), burns of the body and eyes, and radiation syndromes. Near the center of a nuclear explosion, combinations of these disabilities would predominate. The fires and firestorms resulting from the heat impulse would cause numerous cases of asphyxiation due to a shortage of oxygen and toxic concentrations of carbon monoxide. People in the zone of local radioactive fallout would be victims of radiation syndromes.

Indirect effects are those resulting from the destruction of the material structures of society, the disruption of civil services and

Can civil defense save a country against an all-out nuclear
attack? . . . It is impossible.

Lt. Gen. Mikhail Milshtein, USSR Army (Ret.)
Second Congress of International Physicians for the
Prevention of Nuclear War
Cambridge, England, April 3, 1982

the social fabric, the disintegration of the economy, and the
severe disabling of the governmental infrastructure. There also
would be famine and outbreaks of epidemic diseases, a sharp increase in the frequency of other diseases, especially those caused
by infection (tuberculosis, dysentery, and hepatitis, for example),
a rise in various psychiatric disorders among those who survived
the nuclear attack as well as those who, while not directly attacked, nevertheless would be affected by an attack nearby.

Other indirect effects that are difficult to predict could result
from the explosion of multiple nuclear weapons in a war. These
include damaging effects on people and some animals and plants
as a result of the sharp increase in harmful ultraviolet radiation
to the earth's surface because of ozone depletion. Large numbers
of nuclear airbursts with high megatonnage produce enormous
quantities of nitric oxide, which can deplete atmospheric ozone
(see Chapter 19 for a discussion of ozone and ultraviolet). Changes
in temperature in various regions of the earth and global radioactive fallout could also result, having profound effects on human
survival. Other indirect effects, though equally catastrophic, are
not fully understood or perhaps still unknown.

Effects of a nuclear explosion can be further categorized into
early (those occurring up to one month after) and late (those that
appear over the course of months to years). Among the late effects
that are predictable are those secondary to radiation: cataracts of
the crystalline lens of the eye, premature aging syndrome, benign
and malignant tumors, and genetic defects. Although the consequences of nuclear war that are quantifiable, particularly the medical casualties from the blast, heat, and radiation, assume a great
importance, it must be repeatedly stressed that consequences that
are impossible to predict (the societal and ecologic effects) could
be even more catastrophic. This fact must be kept in mind for the
hypothetical nuclear attack that follows.

Effects of a Hypothetical Nuclear Explosion on a Typical City

A one-megaton nuclear explosion instantaneously releases energy on the order of 10^{18} calories. This energy takes the following forms: shock wave, about 50 percent; light (heat) radiation, 30–50 percent; instantaneous ionizing radiation, about 5 percent; and residual radiation, up to 15 percent. Immediately after the explosion, the fireball forms under the influence of the superhigh temperatures. The brilliance of the fireball at a distance of 100 kilometers is 30 times greater than the brilliance of the midday tropical sun. In an airburst, radioactive products are carried away with a rising stream of air to the upper layers of the atmosphere and then to the stratosphere, causing global fallout with a half-life of 1 to 1.5 years. An air explosion in the northern hemisphere would result in 70 percent of the radioactive products falling in that region of the globe. Due to stratospheric transfer, there would be radioactive fallout in the southern hemisphere as well, amounting to approximately one-third of the total quantity of radionuclides produced by the atomic explosions.

Due to intensive dissipation by high winds, global circulation of the airburst product, and the rapid disintegration of short-lived radionuclides on the earth's surface, radioactive fallout from this type of explosion would not present an acute radiation danger for the population. However, global fallout would expose people to small protracted doses of radiation, which could cause delayed radiological consequences.

Surface nuclear explosions present a different scenario. In such cases enormous masses of highly radioactive soil (due to neutron activation) are vaporized and drawn into the fireball, later to resettle as local fallout as the explosion cloud is carried by winds. If 5 percent of the energy released during a surface one-megaton nuclear explosion heated the soil, then 20,000 tons of evaporated soil would be added to the fireball. More than 50 percent of the total radioactive materials from a surface nuclear explosion would settle as local fallout. The remaining radioactive products would be carried away, in the form of highly dispersed particles, to the upper layers of the atmosphere and stratosphere and would settle as global fallout at distances thousands of kilometers away from the explosion.

People subjected to local fallout would be exposed to three types of acute radiation: external, contact, and internal radiation.

A nuclear explosion, taken at a height of approximately 12,000 feet, 50 miles from the detonation site. Two minutes after zero hour the cloud rose to 40,000 feet, the height of 32 Empire State Buildings. (U.S. Air Force.)

The most significant factor of acute radiation would be external gamma radiation from the radioactive materials. Contact beta radiation also would be possible as a result of direct skin contact with fallout. Finally, people in the areas exposed to fallout during the first two to four weeks after the explosion could be subjected to internal irradiation, primarily due to radioactive iodine contained in the milk of milk-producing livestock pastured in fields within the fallout zone. In these instances, the irradiation would most affect children and the fetuses of pregnant women, damaging their thyroids by the uptake of iodine-131 from the contaminated milk. Long-lived radionuclides of strontium and cesium would also expose those people in fallout areas to radiation dangers because of their capacity as internal emitters.

158

We shall consider both a single surface burst and a single air-burst one-megaton nuclear explosion over the center of a hypothetical city with an area of somewhat more than 300 square kilometers (116 square miles), a population of one million people, and an evenly distributed density of 3200 persons per square kilometer (8300 per square mile). The explosions would take place in the daytime, during the summer, with a visibility of 16 kilometers (10 miles). The population would be distributed as follows: 10 percent in open places and 90 percent in brick buildings. The calculated losses for the population near the center of the explosion can be extrapolated from the data at Hiroshima and Nagasaki, with an allowance made for the difference in building materials.

For the case of a surface explosion, along with urban casualties from the three incapacitating forces (shock wave, heat radiation, and instantaneous radiation), there would be acute radiation sickness and delayed radiological consequences in the population exposed to radioactive fallout. These estimates can be quantified. The figure on the next page shows the anticipated numbers of dead and acutely injured following a surface or air nuclear burst on our hypothetical city. Deaths from all factors in an airburst total approximately 300,000; 180,000 would have burns, with 30,000 of this number also having traumatic injuries, while the number of those suffering traumatic injuries alone would reach 200,000. More than one-third of the population, 380,000 people, would require medical care.

In the event of a nuclear explosion over a city, no casualties among the population would have radiation syndrome alone, because the radius of the shock wave's destructive impact and that

---

beta radiation radiation in the form of negatively charged particles, identical to electrons moving at high velocity, which cause cellular damage by contact, either externally on the skin, or internally when ingested or inhaled.

gamma radiation electromagnetic radiation originating in atomic nuclei, physically identical to x-rays. Gamma radiation is emitted both immediately during the nuclear explosion and over time in fallout. Like x-rays, gamma rays are extremely penetrating; living tissue is essentially transparent to gamma rays.

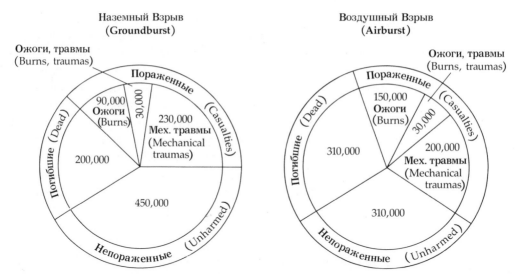

Наземный Взрыв
(Groundburst)

Воздушный Взрыв
(Airburst)

Probable consequences of a nuclear explosion on the population of a hypothetical city of one million people.

of the heat radiation both exceed the radius of the lethal and disabling effects of the nuclear explosion's instantaneous radiation. However, it can be estimated that 3 to 5 percent of the surviving casualties would develop acute radiation syndromes, which would complicate their diagnosis and seriously worsen the prognosis for their burns and traumatic injuries.

Compared to an airburst, a surface nuclear explosion would produce somewhat fewer dead and injured. In contrast to an airburst, however, a surface burst would produce significant residual radiation in the form of local fallout. Assuming a steady wind of 50 km/h (30 mph), and a population density of 50 per square kilometer (225 per square mile) in the local fallout zone, the acute effects of fallout can be calculated. For a radiation protection factor of five for brick buildings (that is, those inside would receive one-fifth of the accumulated radiation dose received by those exposed in the open), the radioactive fallout for a one-megaton surface burst would cause acute radiation sickness over an area of about 4600 square kilometers (assuming all stayed inside the building), with total radiation casualties reaching about 230,000. Of these, 85,000 would die during the first month within an area of approximately 1680 square kilometers. With a protection factor

160

of 1.5, consistent with wooden buildings, acute radiation casualties would number 510,000, in an area of approximately 10,300 square kilometers. For this case, 190,000 would die in an area of 3800 square kilometers. In the zone of fallout, the consumption of cow's or goat's milk contaminated with iodine-131 could result in a concentration of radioactive iodine in the thyroid that delivers an effective dose of radiation ten to one hundred times the gamma dose in the same fallout area. For children, pregnant women, and newborns, this concentration of iodine-131 could be the most significant fallout hazard.

The above discussion has focused on the acute radiation effects. Long-term consequences are also possible to estimate. In the fallout area within which the population would be exposed to an accumulated radiation dose of 100 rads and protected by a factor of 1.5, 30,000 people would die from radiation-caused malignant tumors, and 9000 others could transfer genetic damage to their offspring.

Conclusion

Many of the indirect effects of nuclear war are difficult or impossible to predict and require significantly more study to allow quantitative evaluation. This is particularly true for the effects on the environment. Others, such as the consequences in the breakdown of society, are extremely complex and may not yield to mathematical formulation.

Many of the direct effects of nuclear weapons, by contrast, are predictable and the number of casualties can be estimated. A one-megaton nuclear weapon exploded over a city of one million people would kill or seriously injure approximately 500,000 to 700,000 people, depending on whether it was a surface burst or an airburst. A surface burst would cause an additional 200,000 to 500,000 radiation casualties in the population downwind from the explosion even with some protection from the radiation. Of these, about 90,000 to 190,000 would die. Additional tens of thousands would suffer long-term effects with malignant tumors and genetic damage. These figures, though certain underestimates, illustrate the magnitude of the catastrophe of nuclear war. They offer the strongest possible argument that nuclear wars must never occur.

References

Davidson, G. O. 1960. *The Biological Consequences of General Gamma-Irradiation on Man.* Moscow: Atomizdat.

Ervin, F. R., J. B. Glazier, S. Aronow, D. Nathan, R. Coleman, N. C. Avery, S. Shohet, and C. Leeman. 1962. "Human and Ecologic Effects in Massachusetts of an Assumed Thermonuclear Attack on the United States." In Symposium on the Medical Consequences of Thermonuclear War. *New England Journal of Medicine,* 266:1126–1155.

Ilyin, L. A., et al. 1972. *Radioactive Iodine and the Problem of Radiation Safety.* Moscow: Atomizdat.

Ilyin, L. A., et al. 1975. *Leadership on the Medical Issues of Anti-Radiation Protection.* Edited by A. I. Burnazyan. Moscow: "Medecine."

National Academy of Sciences. 1963. "Damage to Livestock from Radioactive Fallout in the Event of Nuclear War." Washington, D.C.: NAS-NRC, Publication 1078.

Petrov, R. V., et al. 1963. *Protection from Radioactive Fallout.* Moscow: Medgiz.

Radiation Protection. 1978. Moscow: Atomizdat, MKRZ Publication 26.

12

The Possible Consequences
of a Nuclear Attack on London

Andrew Haines, M.B., B.S.

Although it is not possible to be certain where missiles would land in the event of a nuclear attack on any country, some places are more likely targets than others. The NATO "Square Leg" exercise, which took place in 1980, provides a credible pattern of missile strikes, which can be used as a model to assess the possible consequences of a nuclear attack on Great Britain (Campbell, 1980).

According to this scenario, a nuclear attack would follow a period of tension during which the Cabinet would approve the suspension of Parliament and the assumption of emergency powers. Finally, war would be declared. Panic buying would follow, and many people would attempt to leave major population centers and areas close to military bases, despite government exhortations to remain where they were. All fourteen major roads from London would be designated "Essential Service Routes" for government traffic only. It is not clear what sanctions would be used against unauthorized persons attempting to use these routes; it seems unlikely, however, that peaceful measures would be successful against large numbers of frightened people.

The assistance of Professor J. Rotblat, Jane MacAuley, Alison MacFarlane, and Duncan Campbell is gratefully acknowledged.

Table 1. Description of nuclear bursts within Greater London.

Target	Megatons	Height of Burst
Heathrow	1	Surface
Heathrow	2	12,000 feet
Croydon	3	Surface
Brentford	2	Surface
Potters Bar	3	14,000 feet

Table 2. Blast casualties.

	Millions of People
Total Population, Greater London	7.2*
Immediate blast fatalities	1.1+
Injuries due to blast	2.4–2.9
Uninjured by blast	3.2–3.7

*Private households only.

In the midst of this background of increasing panic, the nuclear attack would occur. Table 1 shows the targets and the size and height of the nuclear bursts for the bombs landing within Greater London. In addition, several other bombs would land outside this area, some of which would cause added, although light, blast damage to London. Adverse effects outside Greater London are not included in this report.

By calculating the overpressures (measured in pounds per square inch) generated by the bombs and by using the 1971 census statistics available from the Office of Population Censuses and Surveys, overpressure rings can be superimposed on a map of Greater London and blast casualties from the attack can be estimated. The overpressure rings are shown in the figure opposite, and injuries and fatalities are given in Table 2. The range in the number of injuries is due to the overlapping nature of the blasts. The lower estimate assumes no additive effect from overlapping blasts on the number of injuries; the higher estimate assumes that there are additive effects, which in practice probably would be more likely. Even the higher figures might be underestimates be-

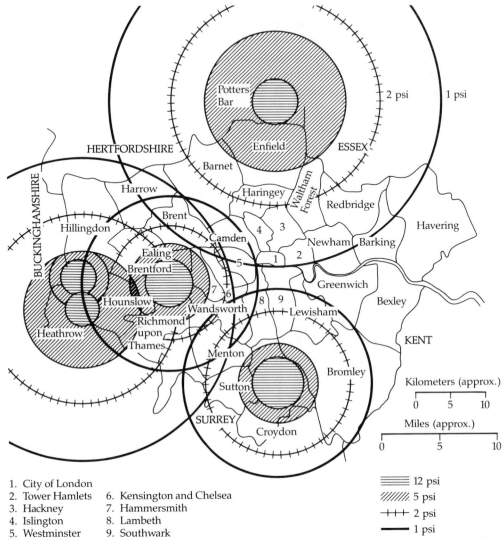

Overpressure rings for nuclear attack on Greater London.

1. City of London
2. Tower Hamlets   6. Kensington and Chelsea
3. Hackney   7. Hammersmith
4. Islington   8. Lambeth
5. Westminster   9. Southwark

cause the combined effects might be greater than the sum of those occurring separately.

Blast injuries would occur mainly from the collapse of buildings and from collisions between people caught in the open and solid objects. Injuries from flying glass would be extremely common, with serious cases extending out to the 2-psi overpressure

165

As a military man who has given half a century of active service, I say in all sincerity that the nuclear arms race has no military purpose. Wars cannot be fought with nuclear weapons. Their existence only adds to our perils because of the illusions which they have generated.

Lord Mountbatten
Former Chief of the Defense Staff, United Kingdom
Former Chairman of the NATO Military Committee
"A Military Commander Surveys the Nuclear Arms Race,"
International Security, Winter 1979/1980

ring. Eardrums rupture at overpressures above 5 psi, creating additional problems for the survivors.

In Hiroshima and Nagasaki there were relatively few serious mechanical injuries among those who survived the initial blast. This low incidence was thought to be due to the firestorm, which killed those with severe injuries who were unable to escape. Western cities may be less likely to support a firestorm, so that possibly a significant proportion of initial survivors would suffer from severe trauma.

Parts of crumbled buildings, downed power and telephone lines, destroyed vehicles and other structures shattered by the blast would block most of the streets within the 5-psi ring and make evacuation of the injured difficult, if not impossible. Even out to the 2-psi ring, debris on the streets would interfere with movement.

The number of people sustaining burns in a major attack on an urban center is difficult to predict with any accuracy. It depends on the percent of the population in the open at the time of the blast, visibility, clothing worn, and even skin color (darker skins absorb more heat). The approximate number of major burns that would occur in two cases (1 percent and 25 percent of the population in the open) is given in Table 3. Only flash burns, the direct result of thermal radiation from the explosion, are included in Table 3.

Because of the difficulty in predicting the extent of the fires, flame burns, which occur as a consequence of the fires ignited by the explosion, are not estimated. As a result the numbers given in

Table 3. Flash burns in blast survivors (12-mile visibility).

Overpressure Ring (psi)	Number of People Burned		Types of Burns
	1% Outdoors	25% Outdoors	
1–2	Some	Some	Second degree Many first degree
2–5	28,000	700,000	Mainly second degree Many third degree Some first degree
5–12	5000	130,000	Mainly third degree

Table 3 are marked underestimates. (For a more complete discussion of flash and flame burns, see Chapter 7.)

In a situation such as that envisaged in "Square Leg," when war had already been declared, it is quite possible that large numbers of people would attempt to flee London, so that a greater percentage of the population might be caught in the open by the thermal flash. It is most unlikely that fire fighting services could make any impact on the spread of fires, because of impaired mobility, the size and number of the fires, and the lack of water pressure. In addition, the high levels of residual radioactivity in the immediate post-attack period would make lethal any exposure in the open.

The total number of injuries would depend on the number of blast injuries also suffering burns. For instance, if all the burns occurred in people who were injured by the blast, the total would be 2.4 to 2.9 million (Table 2). However, if all burns occurred in those who were not injured by blast, the total would be approximately 3.2 to 3.7 million (assuming 25 percent of the population were in the open at the time of the attack). A reasonable estimate of the total number of injuries probably lies somewhere between these extremes.

Flashblindness in those looking toward the blast would add to the horror of the immediate post-blast situation. Although the blindness would, in almost all cases, be temporary, those affected might not be aware of this. They also would be particularly vulnerable to further injury by fire or falling debris. Persistent visual

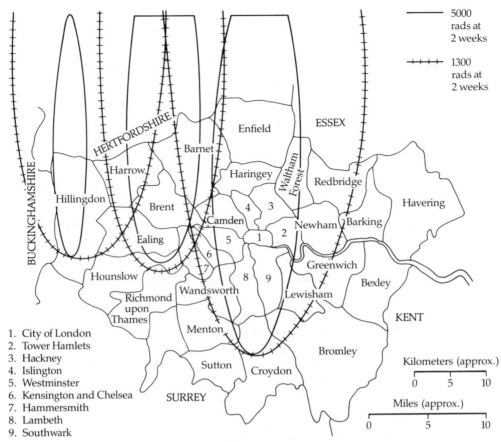

1. City of London
2. Tower Hamlets
3. Hackney
4. Islington
5. Westminster
6. Kensington and Chelsea
7. Hammersmith
8. Lambeth
9. Southwark

Fallout patterns for nuclear attack on Greater London. Winds are southerly. The ellipses outline areas for accumulated radiation doses of 5000 and 1300 rads for a two-week period. Average wind velocity is 15 mph.

field defects due to retinal burns could occur many miles away from the blast, particularly if the flash occurred on a dark night when the pupil was dilated and the retina relatively unprotected.

If the winds were steady from the south, fallout from the surface burst bombs would spread across central and northern sections of London. If the wind direction was from the west, a more common condition, all of central and eastern London would be covered. An idealized fallout pattern from the ground bursts, assuming southerly winds at 15 mph, is shown in the figure above for two-week accumulated doses of 5000 and 1300 rads. The dis-

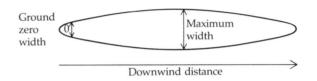

Downwind distance

Accumulated dose at two weeks (rads)	Downwind distance (miles)	Downwind distance (km)	Maximum width (miles)	Maximum width (km)	Ground zero width (miles)	Ground zero width (km)
6000	21	34	2.9	4.7	1.4	2.3
1500	40	65	6.8	11.0	3.1	5.0
300	100	162	12.4	20.0	5.5	8.9
75	195	321	23.9	38.6	7.1	11.4

Fallout distances from a one-megaton surface burst, assuming 50 percent fission and 15-mph wind velocity.

tances fallout from a one-megaton bomb would reach are given in the figure above. Even with single surface atomic explosions, the actual fallout patterns are highly irregular, and the idealized patterns are, at best, extremely rough guesses; with multiple explosions, the fallout patterns are even more unpredictable.

Multiple bursts over a small area, for example, would be likely to cause local wind turbulence, which would have marked effects on fallout deposition. It seems unlikely, therefore, that any worthwhile monitoring of fallout levels could be performed in these circumstances unless the monitoring equipment and personnel were protected from the blast effects and, in the case of the personnel, from the fallout itself. Moreover, the monitoring stations would have to be numerous and mobile because of the unpredictability of the fallout patterns—a virtual impossibility, given the situation. Many houses in the areas covered by fallout would have been damaged by blast, and perhaps by fire, and would offer little protection against fallout.

To survive a dose of 5000 rads accumulated over a two-week period, the population surviving the initial effects of the bombs would probably have to be in effective shelters giving a protec-

169

tion factor of more than 10, that is, shelters shielding out the radiation so that occupants are exposed to one-tenth or less of the outside levels of radiation. Most basement rooms would require considerable volumes of earth piled against exposed walls and windows to offer radiation protection factors of more than 10. However, only 3.5 percent of dwellings within the Greater London boundaries have basements, and about one-third of these are basement flats to which other households in the buildings might not have access.

More than 40 percent of all dwellings within the Greater London boundaries are flats, and many of these, particularly those on the upper floors of block rows or houses, offer poor protection from blast and fallout. To construct an effective fallout shelter within a house, it would be necessary to close windows and doorways with brick and to move several cubic meters of earth (1 cubic meter (about 1 cubic yard) of earth weighs approximately 1465 kilograms or more than $1\frac{1}{2}$ tons). Even if the materials were available, these tasks would be beyond the capability of most households. Moreover, many people would experience pre-attack panic and be unable to plan or work effectively.

The proportion of survivors of the initial effects of the bombs who would find shelter in undamaged houses with properly constructed fallout shelters would be very small. Many initial survivors would therefore perish from the effects of fallout within a few weeks after the attack. Those in shelters would not know the dose of radiation they received. Since the early symptoms of radiation sickness are nonspecific, it would be impossible to know at first whether a person was suffering from radiation sickness or from vomiting due to stress or gastroenteritis. This uncertainty would be likely to lead to further widespread panic and demoralization among both early survivors and medical workers treating the casualties.

The problems that would confront the medical care system even if it were fully intact would be overwhelming, considering the enormous number of casualties. But, in fact, the number of

---

rad unit of radiation; 450 rads in a short period of time is a lethal dose for approximately 50 percent of healthy adults exposed.

Hiroshima railway station, 2 miles from the hypocenter. The bomb was dropped at commuting time. The blast blew out the windows, and fire gutted the building. The upper floors of buildings in target areas would offer poor protection against blast and fallout. (Shunkichi Kikuchi, Hiroshima–Nagasaki Publishing Committee.)

functioning medical personnel and facilities would be drastically reduced. For example, there are 8000 hospital doctors and 3500 general practitioners for a total of 11,500 in Greater London. Assuming that they survived uninjured in the same proportion as the population as a whole, there would be 4000 to 6000 doctors able to work following the "Square Leg" nuclear attack. If we then assume that all patients somehow reached these surviving doctors, overcoming the problems of clogged access routes and residual radiation, then there would be approximately 400–900 injuries including 6–175 major burns per doctor. If each doctor worked 18 hours per day and allowed 20 minutes per patient, it would take 7–17 days to provide even the most minimal medical care, by which time many would have died.

In reality, the situation would be considerably worse than this. Many injured would never see doctors. Supplies of medicine and blood would be limited. Treatment would consist of simple surgical procedures, often without anesthetic. Lethal levels of fallout would preclude treatment of the injured over large areas, unless the medical staff were prepared to become fatalities themselves. To compound the magnitude of the disaster still further, most of the hospitals in Greater London would be destroyed. Of a total of approximately 60,000 hospital beds, fewer than 24,000 would remain after "Square Leg," of which only 14,000 would be suitable for acute care.

The figures in the preceding pages attempt to quantify the devastation and massive destruction of Greater London and its population following a nuclear attack. Many of those who survived the initial effects would die later of radiation sickness, or lack of water and food. Proposed civil defense measures cannot provide adequate protection against such a holocaust, and medical facilities would have little to offer.

Reference

Campbell, Duncan. 1980. "World War III: An Exclusive Preview." *New Statesman*, March 10, pp. 5–6.

13

Fatalities from a One-Megaton Explosion over Tokyo

Naomi Shohno, Ph.D.,
and Tadatoshi Akiba, Ph.D.

In the following short article we attempt to estimate the number of deaths in the aftermath of a one-megaton nuclear airburst 1.8 kilometers over Tokyo. The estimate can be based on figures from Hiroshima, since both cities have similar flat terrains.

The majority of deaths occurred at Hiroshima during the four-month period (August to December 1945) after the bomb was dropped. And more than 60,000 *hibakusha* (those exposed to the bomb) died during the five years after this initial period had elapsed—from December 1945 to October 1950. Deaths traced to the persistent effects of radiation (primarily from leukemia and solid tissue cancers) occurred even after 1950 and continue to occur. These intermediate and long-term deaths at Hiroshima cannot be used as a model for Tokyo, however, because of a number of complicating factors. The four-month totals, then, form the basis for the calculations made below.

The exact figures at Hiroshima cannot be obtained, since almost all administrative agencies and official documents were destroyed by the explosion. From the best estimates available, the population at Hiroshima on August 6, 1945, was between 340,000 and 350,000, including about 40,000 military personnel. Of these, approximately 140,000 people had died as a result of the bomb by December 1945. How many military personnel were among them is not known.

The mushroom cloud of the Hiroshima bomb. August 6, 1945. (Hiroshima–Nagasaki Publishing Committee, U.S. Army returned materials.)

Table 1 summarizes the four-month mortality rate at Hiroshima. The rates are the percentage of the total population who died within each ring at various distances from the hypocenter. Within the ring between 0.5 and 1.0 kilometer from the hypocenter, for example, 83 percent of the total population died. Within the total area enclosed by the 2-kilometer radius (approximately 12.5 square kilometers or almost 5 square miles), the total mortality rate was approximately 60 percent, corresponding to 125,000 deaths out of a total population in that area of 200,000 to 210,000. In this same zone, all buildings and structures were entirely destroyed or burned by blast and fire.

The population of Tokyo is approximately 10 million, with an average population density significantly greater than that of Hiroshima (16,000 to 16,800 people per square kilometer) at the time of the bombing. Although it is difficult to determine with accuracy the population density of Tokyo because of fluctuations influenced by the time of day and the area of the city, for the purpose of our calculation an approximate figure of 20,000 people

174

Table 1.   Four-month mortality rate at Hiroshima by distance from the hypocenter.

Distance from Hypocenter (km)	Mortality Rate (%)
0–0.5	97
0.5–1.0	83
1.0–1.5	52
1.5–2.0	22
2.0–3.0	4
3.0–5.0	2

per square kilometer will be used. This figure probably is an underestimate.

We must then determine the distance from the hypocenter of a one-megaton explosion, within which the blast casualties would be comparable to those inside the 2-kilometer radius at Hiroshima. The yield of the Hiroshima bomb was 12.5 kilotons; a one-megaton weapon would yield 80 times as much energy. By using a simple calculation, we can estimate that at 8.8 kilometers from the hypocenter of a one-megaton blast, the overpressures and wind velocities would be approximately the same as those experienced at 2 kilometers in Hiroshima. We are able, therefore, to calculate the blast fatalities for Tokyo within the 8.8-kilometer circle by extrapolating from the 2-kilometer circle at Hiroshima. Although the blast effects at 8.8 kilometers for a one-megaton airburst would be approximately the same as at 2 kilometers for a 12.5-kiloton airburst, the thermal effects are not. Large nuclear explosions generate a greater amount of thermal energy proportional to blast energy. A one-megaton airburst, for example, produces 17.5 calories per square centimeter at 8.8 kilometers from the hypocenter, compared to 6.5 calories per square centimeter at 2 kilometers from the hypocenter at Hiroshima. Seventeen and one-half calories is intense enough to ignite almost all combustible materials (clothes, furniture, wooden structures) and also to char human skin (12 cal/cm^2 is sufficient to cause third-degree burns on human skin).

The population of Tokyo within the 8.8-kilometer radius can be estimated at 4,866,000, using the density of 20,000 people per square kilometer. Assuming a Hiroshima mortality rate of 60 per-

It was at that moment . . . the sound . . . the lights out . . . all was dark . . . . How I got out, I don't know . . . the sky was lost in half-light with smoke . . . like an eclipse . . . . The window frames began to burn; soon every window was aflame and then all the inside of the building . . . . There were eight of us there . . . we sat on the stone steps . . . . A woman said one eye was going blind; one man said he was feeling bad; another's head was aching . . . . The fire spread furiously and I could feel the intense heat . . . smoke whirled up, filling the air, hurling sheets of burning zinc and flying cinders. Looking up, we could see the danger. Smoke was blinding us, bringing tears to our eyes, and almost stifling us . . . we took up zinc sheets to protect our heads. Suddenly the water of the river rose like a water-spout sucked up by the wind, and fell in a shower to leeward. The force of the fires grew in violence, and sparks and smoke from across the river smothered us . . . . We dipped our zinc sheets in the river and put them over us — again and again we did it and barely managed to escape . . . .

Eizo Nomura, Hiroshima
Hiroshima–Nagasaki: A Pictorial Record of the Atomic Destruction

cent, we arrive at a total of 2,919,000 fatalities. This figure is likely to be a considerable underestimate of the fatalities during the first four months at Tokyo. First, the Hiroshima mortality rate of 60 percent is based on minimum figures. Second, the 60 percent figure used for the 8.8-kilometer radius in Tokyo seriously underestimates the effect of the thermal energy at that distance that would be likely to cause additional deaths from burns. The 60 percent mortality rate may extend far beyond the 8.8-kilometer radius, but it is difficult to estimate how far. For these reasons, we believe that the mortality figure for Tokyo could be as high as three or four million people.

Summary

If a one-megaton nuclear bomb were to explode 1.8 kilometers above the center of Tokyo during daytime, all structures within an 8.8-kilometer radius from the hypocenter would be destroyed by blast and fires. Mortality figures in the area for the four-month period after the explosion, from traumatic injuries and severe burns, would be greater than 60 percent of the total population of 4,866,000 in that area, or approximately 3,000,000 people, and could reach 4,000,000. Long-term effects from the explosion over the ensuing months and years would add hundreds of thousands, and perhaps more than a million, additional deaths to this total.

SECTION IV

Nuclear War, 1980's:
The Medical Response

Available data reveal that there is no adequate medical
response to a nuclear holocaust. In targeted areas,
millions would perish outright, including medical and
health care personnel. Additional millions would suffer
severe injury, including massive burns and exposure to
toxic levels of radiation, without benefit of even
minimal medical care. Medical and hospital facilities
and other resources would likewise have been destroyed.

American Medical Association
Report of the Board of Trustees
December 1981

14

The Immediate Medical Response

Alfred Gellhorn, M.D.,
and Penny Janeway

The role of the physician in the post-nuclear-attack period is difficult to contemplate in light of the known devastating effects of
nuclear weapons explosions. But as the buildup of nuclear arsenals accelerates, so does the possibility of nuclear war, and it is
unacceptable to avoid examining the demands it would place on
the medical system. The examination that follows reveals the impossibility of providing adequate medical care for the injured millions and demonstrates, perhaps more effectively than anything
else, that nuclear wars cannot be won.

It is easier to think about the medical needs of the injured,
whatever the enormity of their number, if we divide them into
three categories: traumas, burns, and radiation injuries. It should
be kept in mind that many people would have more than one
kind of injury, with each complicating the effects of the other.

To understand better the kinds of medical treatment required
for the management of these nuclear injuries, let us look at actual
cases of trauma, burns, and radiation exposure, which can serve
as models, with the difference, however, that these cases were
treated under normal conditions and were not caused by nuclear
weapons. (Burns and radiation injuries are discussed more fully in
Chapters 15 and 17.)

Blast Injuries

On April 16, 1947, a fire was discovered on a freighter moored at a slip in the harbor of Texas City, Texas. The ship's cargo, consisting of ammonium nitrate fertilizer, a large amount of twine balls, and small arms ammunition, caught fire and exploded, destroying the freighter with a tremendous blast and hurling fragments of steel in all directions. Secondary fires were started by red-hot metal particles and balls of burning twine. The force of the explosion produced a tidal wave 10 feet high.

Five hundred and sixty people were killed outright, and 800 patients were hospitalized with serious injuries. The deaths were due to a variety of blast injuries, dismemberment from flying debris, and burns from the secondary fire in an adjacent chemical plant.

Medical facilities in Texas City were unable to cope with the emergency, so the severely wounded had to be transported to Galveston and Houston. There were certain common features among the 800 casualties admitted for hospitalization. Many of the injured had a large number of point-like lacerations concentrated on the head and extremities, usually from imbedded glass particles. Simple and compound fractures of facial bones and extremities were also very common. One-third of the patients were deaf in either one or both ears from rupture of the tympanic membrane. People with extensive burns or massive wounds were conspicuously absent. They had not survived.

Triage consisted of starting plasma and antibiotics and sending hemorrhaging patients for immediate surgery. Patients with simple fractures were sent to the cast room. Those with multiple lacerations (without perforation of the abdominal wall or thorax) were referred to minor surgery rooms for cleansing, removal of dead tissue, and suturing of wounds. Tetanus antitoxin was given to all patients. Penicillin, oxygen, plasma, blood, intravenous salt solutions, and pain killers were extensively used. Only seven cases of gas gangrene developed, but of these three were fatal.

Ten operating teams with three or four surgeons in each team worked in shifts around the clock for 48 hours. Seventy-two physicians provided care in the wards, and numerous specialists in orthopedics, ophthalmology, radiology, and anesthesiology were available. Red Cross and volunteer nurses were present in large

Passengers thrown out of a streetcar, near the stone walls, when the bomb exploded. They were killed instantly by the blast and heat. The parts that look black are actually deep red, the effect of thermal burns. Nagasaki, August 10, 1945. (Yosuke Yamahata, Hiroshima–Nagasaki Publishing Committee.)

numbers, as well as the regular and student nurses. In the first week following hospitalization, there were only 13 deaths.

In this example we see the medical care system of a medium-sized American city operating with full emergency capability. Even so, the assistance of two other cities was required to care for these 800 patients. This is a small fraction of the number of blast victims (including among them many medical personnel) to be expected from even a single nuclear explosion. And any nuclear explosion in any city will almost certainly have destroyed the medical facilities as well. And given the extreme unlikelihood of a limited nuclear war, there would be no outside cities ready to assist.

A Burn Accident

A $4\frac{1}{2}$-year-old girl was alone in her home when an explosion and fire occurred in the kitchen. Terrified, she ran to another room and hid. By the time firemen finally found and rescued her, she had been severely burned. She was rushed to a large American children's hospital where it was found that 40 percent of her body surface was burned, including her face, scalp, chest wall, anterior abdominal wall, and upper and lower extremities; there were deep third-degree burns on her feet.

Resuscitation measures, including intravenous plasma and a specialized salt solution, were started immediately. Careful cleaning of the burned areas, shaving of her scalp hair and application of an antibacterial sulfa dressing followed. Topical antibiotics were applied every two hours, with special attention to her burned eyelids and conjunctivae.

Six days after admission, the child underwent the first of twelve extensive surgical procedures, which were to extend over a year. The first procedure was confined to removal of dead tissue in her wounds. One month later, she had local and systemic infections from which both bacterial and fungal microorganisms were cultured. Antibiotics specifically directed against the microbial invaders successfully controlled the infections.

Surgical removal of dead tissue was performed on four occasions, including amputation of fingers and toes. Later three grafting procedures and two plastic surgery operations were per-

184

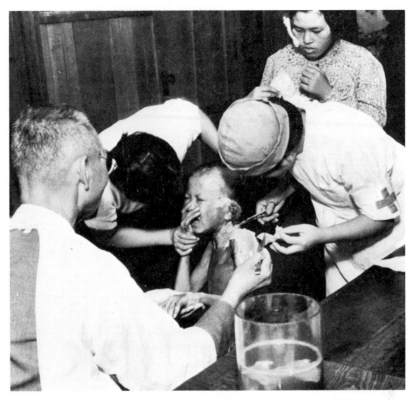

A child crying from the pain of having a gauze dressing changed. He suffered third-degree burns that exposed the bone. Nagasaki, early September, 1945. (Yasuo Tomishige, Hiroshima–Nagasaki Publishing Committee.)

formed. The Hand Service projected that surgery extending over years would be required for restitution of the function of her hands and legs.

Over the succeeding $9\frac{1}{2}$ months she had multiple skin grafts for nonhealing burns, for release of burn contractions, for excision of burn-induced skin tumors, and for delicate plastic surgery of the eyelids and eyebrows.

The overwhelming psychological impact of this accident and the aftermath of prolonged and painful medical care has not been assessed in this young child, but the expenditure of human and material resources during the first year alone could be measured in the hundreds of thousands of dollars.

This report of a single severe burn case and the medical response to it suggests the enormous magnitude of the effort that would have to be made to meet the needs of burn casualties caused by a nuclear weapon explosion. The estimates of burn casualties that would result from a single one-megaton weapon detonated over one city, Detroit, for example, range from 12,000 to more than 300,000.

Burn cases place a very great strain on medical personnel. The British Army Operational Research Group, utilizing evidence from wartime England, estimated an average time of 52 minutes for three persons to simply dress a burned hand. They calculated that the peacetime requirement for treating 34,000 serious burn cases would be 170,000 health professionals and 8000 tons of supplies.

Sublethal Radiation Exposure

A 38-year-old man was accidentally exposed to a radioactive cobalt-60 sample at a commercial laboratory that prepared radioactive products for medical therapy and diagnosis. The calculated exposure was 200 rads.

Two hours after the accident, the patient had nausea and was vomiting severely. He was treated with medication to stop the

---

normal values
  platelets   200,000–500,000 per cubic millimeter of blood.
  white blood cells   5000–10,000 per cubic millimeter of blood.
  hemoglobin (males)   14–18 gram-percent (14–18 grams of hemoglobin in 100 cubic centimeters of whole blood).

rad   unit of radiation; 450 rads in a short period of time is a lethal dose for approximately 50 percent of healthy adults exposed.

triage   classification of a large number of casualties into three groups: those who will survive without any medical help, those who will die no matter what treatment they receive, and the priority group of those who will survive only if they receive medical treatment.

---

vomiting, but because of increasing malaise, vomiting, and diarrhea, he was admitted to a hospital 15 hours after the accident. He was immediately placed in a reverse isolation facility (designed to prevent his exposure to infectious agents).

On the 12th day his white blood cell count had fallen from 5000 to 1000, while his hemoglobin concentration and platelet level were unchanged. He was started on antibiotics to achieve bowel sterilization. One month after the accident his platelet count was less than 10,000, his white blood cell count less than 500, and his hemoglobin had fallen from 16 gram-percent to 8 gram-percent. The patient had fever and had lost scalp and facial hair. On additional antibiotics and other supportive measures, his bone marrow function returned and the patient was discharged 47 days after the accident, although still with a depressed white blood cell count.

During the hospitalization the special requirements included a reverse isolation unit, specialized staff to care for the patient in the unit, and extensive laboratory monitoring of peripheral blood counts and of blood cultures.

A very conservative estimate of the dollar cost of his treatment would be $22,000, but similar care involving such personnel and facilities would be impossible in most cities if the number of radiation accident victims exceeded a hundred (Eiseman and Bound, 1978).

Lethal Radiation Exposure

Late in the afternoon of July 24, 1964, a 37-year-old married father of nine was pouring a mixture of radioactive isotopes including uranium-235 into a tank as part of a standard process in a uranium-235 recovery plant. Near the completion of this step a critical volume and geometry was attained, and a limited nuclear reaction occurred. The patient recalled a flash of light and was hurled across the room, stunned but conscious. He ran from the building to an emergency hut 200 yards away, tearing his clothes off as he ran. Almost at once he complained of abdominal pain and headache, vomited, and was incontinent of bloody diarrhea. He was wrapped in warm blankets and taken to a large general hospital one hour and 43 minutes after the accident.

187

Complaining of severe abdominal cramping, headache, chills, and thirst, he was taken at once to an isolated section of the emergency service. He was perspiring heavily and incontinent of diarrheal stool. Except for abdominal rigidity, the physical examination was within normal limits. Because of restlessness and pain, morphine and an antihistamine were given and plasma was begun intravenously.

A survey of the body surface was made for gamma ray emission, and the levels of radiation measured were hazardous to attendants. A thorough bath on a plastic sheet substantially reduced the readings.

Two hours after admission his blood pressure began to fall and for four hours it was at shock levels. Drugs to raise the blood pressure and antibiotics were given intravenously, and other antibiotics by gastric tube. Eight hours after admission his left hand and forearm, which had held the container of uranium-235, became swollen and red. There was swelling around his eyes, which were bloodshot.

Efforts to support the blood pressure with plasma and drugs were unsuccessful, and kidney function began to fail. The left forearm progressively swelled and the swelling extended upward, vision deteriorated, and the patient became disoriented and hyperactive. Blood pressure was no longer obtainable. The patient died 49 hours after the accident.

An estimation of the gamma and neutron exposure was a total whole-body dose of 8800 rads, or between ten and twenty times the average lethal dose. Even after thoroughly bathing the patient, those attendants who were exposed to his body at a distance of 2 feet for a period of nine hours received the maximum allowable radiation exposure for one week. All vomitus, feces, and urine were contaminated with radioactive elements, presenting significant problems of disposal. All those attending the patient were required to wear caps, gowns, masks, plastic gloves, and grocery sacks tied over their shoes.

This serious nuclear accident not only had a tragic outcome for the victim but also generated health hazards for health professionals and administrative problems related to radiation contamination. The involvement of health professionals, auxiliary personnel, equipment, supplies, and an isolation room presented a significant issue to the hospital, despite the brevity of the patient's terminal course. As stated by the author of the medical report, "One shudders to imagine the difficulties generated by an accident involving major exposure of several or more persons" (Eiseman and Bound, 1978).

Implication of Combined Injuries

Reviewing these examples of blast injuries, burns, and radiation exposure places in perspective the scope of the medical care problems for survivors of a nuclear attack. Unlike the distinct types of traumatic, thermal, or radiation injuries usually seen in civilian accidents, however, the casualties from a nuclear weapons explosion would have combinations of these.

Surgery for severe burns or a traumatic injury must take place within 96 hours after the injury in order to avoid tissue breakdown and infection. However, a person who suffered radiation injury, in addition to severe burns or a fracture, could not be treated until the lengthy process of recovery from the radiation injury was well under way, because proper healing would not occur until then. Bone repair and the regeneration of bone marrow and soft tissue following radiation injury require weeks or even months. For a minimum of 12 weeks, then, only supportive care for those with combined radiation and traumatic injuries could be provided. The need for surgery to prevent infection would conflict with the body's need to heal first, creating an impossible dilemma for the physician.

This combination of burns or traumatic injury with radiation injury would be common among nuclear war survivors. Sterile environments would be necessary but unavailable. Those with major traumatic injuries and burns would become fatalities; minor injuries, because of complications, would become major ones. The scarce hospital beds would be filled with injured patients awaiting recovery from radiation sickness.

Surviving Physicians

Having looked at individual cases, treated in peacetime with an intact medical system, let us consider the casualty figures projected in the event of nuclear attack. Taking metropolitan Wash-

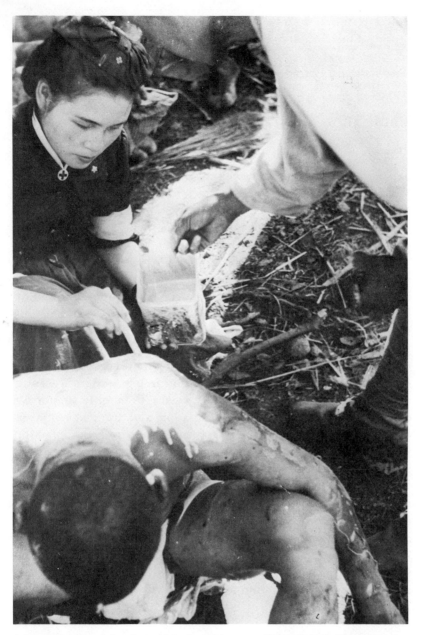

A hospital nurse giving first aid for burns in unsterile conditions. Nagasaki, at the railway station, August 10, 1945. (Yosuke Yamahata, Hiroshima–Nagasaki Publishing Committee.)

190

Our knowledge and credentials as physicians do not, of course, permit us to discuss security issues with expertise. However, if our political and military leaders have based strategic planning on mistaken assumptions concerning the medical aspects of a nuclear war, we do have a responsibility. We must inform them and the American people of the full-blown clinical picture that would follow a nuclear attack and the impotence of the medical community to offer a meaningful response. If we remain silent, we risk betraying ourselves and our nation.

Howard H. Hiatt, M.D.
Dean, Harvard School of Public Health
Journal of the American Medical Association,
November 21, 1980

ington, D.C., with a population of 2.5 million as an example: a one-megaton airburst would leave 600,000 dead and 800,000 injured. Using the projections of killed and wounded for the general population, we can extrapolate that of the 6000 physicians in the area, 1500 would be killed at once and 2000 would be seriously wounded. This would leave 2500 doctors to treat 800,000 people. It would take five 16-hour workdays for those remaining doctors to visit each patient if they spent only 15 minutes with each, and if their patients were somehow centralized so that no time was lost between consultations. Of course, almost none of the injured could be diagnosed and treated in 15 minutes; most would require hours, days, weeks, and months of continuous medical attention, as the prior examples illustrate.

Casualties among physicians are likely to be higher than the rates for the general population in a direct hit on a city, because physicians tend to be close to or in hospitals, which usually are located near the centers of cities.

Availability of Facilities and Supplies

What facilities would remain in which the few surviving physicians could treat the seriously injured? The projection of damages to Detroit under a one-megaton attack in the study by the Con-

The interior of the operating room at Nagasaki University Hospital, 2200 feet from ground zero. Fire has consumed the floor, the balcony, and all seats and has distorted the metal railings and pipe. (Office of U.S. Strategic Bombing Survey.)

gressional Office of Technology Assessment (OTA) calculated that 55 percent of the metropolitan area's 18,000 hospital beds would be inside the ring in which the overpressure would reach 5 psi or more. Lightly constructed commercial buildings and typical residences would be destroyed at 5 psi; heavier construction would be severely damaged. The OTA estimated that 5000 beds would remain, enough for 1 percent of the injured. Suppose, for the sake of this discussion (although it would almost certainly *not* be the case), that Detroit was the sole American city attacked, and therefore the injured would have access to all the hospital beds in the United States, which in 1977 numbered 1,407,000 (remember that most of these beds would be occupied). The number of injured for Detroit alone has been estimated, depending on the type of attack, at between 420,000 and 1,360,000.

Another obstacle to obtaining usable medical supplies is the short shelf-life of medicines. Penicillin G tablets lose their effectiveness after five years. Streptomycin in dry form has a four-year dating, tetracycline two years, and tetanus antitoxin three years.

Blood supplies are not kept in sufficient quantity in any city to deal with mass emergencies. For the city of Boston, for example, the Red Cross whole blood inventory (all types) on a random day in January 1981 was 4619 units; for all of New England the total was 9272 units. Reports from the Vietnam War, from the Yom Kippur War between Israel and its Arab neighbors, and from records of present blood use in civilian hospitals indicate that patients requiring whole blood transfusion each use an average of five units. There is, therefore, enough blood on hand in New England to transfuse about 2000 injured people (assuming no difficulties in cross-matching), or in Boston about 1000 people under normal circumstances. The Boston supply most likely would be destroyed in a nuclear attack on Boston. Moreover, the difficulty in obtaining the transfusion bags and needles, stored for New England in a suburb 10 miles from Boston, would be extreme. An average of 1000 units of whole blood is collected in the New England region each day, enough to transfuse 200 people. It is doubtful, however, that within the 24 hours following a nuclear attack—a crucial time period for the treatment of blast injuries—this rate of production could be maintained.

The total production of whole blood in the 58 Red Cross regions in the United States in all of 1979 was 5,214,901 units. In the event of a nuclear attack upon one medium-sized city, the entire U.S. blood supply for a year could be required within the first 24 hours.

Further Complicating Factors

A superficial knowledge of nuclear weapons effects and a limited exercise of the imagination are all that is necessary to conclude that there is an almost infinite range of factors that would hinder efforts toward post-attack medical recovery. Danger from radiation exposure would limit rescue attempts everywhere, preventing movement of health workers and patients in the open. Fires would continue to smolder and burn in all parts of the target area, keeping temperatures at high levels, further blocking transportation. Rubble would fill streets and make access to patients and

Impassable streets 1500 feet from the hypocenter at Hiroshima. August 7, 1945. (Mitsugu Kishida, Hiroshima–Nagasaki Publishing Committee.)

care centers often impossible. The OTA reports that the depths of debris would depend on both the heights of buildings and the spacing between them: "typical depths might range from tens of feet in the downtown area where buildings are 10 to 20 stories high, down to several inches where buildings are lower and streets are broader." The report goes on: "In this band [5 psi] blast damage alone will destroy all automobiles . . . few vehicles will have been sufficiently protected from debris to remain useful."

Secondary effects as well would complicate medical care—such as lack of electricity, communication, and clean food and water. Sanitation and the disposing of the hundreds of thousands of dead also would be significant problems.

Consider the immediate difficulties the lack of electricity would create. The OTA study of a nuclear attack on Detroit specified that the explosion occurred at night. After the initial flash of light, the world might be dark for all. Although Detroit's electrical power plants are located at the city's periphery and might escape significant direct damage, the electrical grid might well be lost by destruction of the main trunk lines to the city, or by the surge from an electromagnetic pulse. Thus, at the moment of maximum panic, no one would be able to see. The moans and screams of the seriously injured who could profit only from immediate medical attention would be indistinguishable from those of the very young and the panicked. It would be essentially impossible to judge where needs were greatest, where dangerous obstacles lay, and where effective help could be given. Although civil defense booklets advise stockpiling flashlights in shelters, batteries, like medicines, have limited shelf-lives, and besides, many people would be far from prepared shelters. For most of those in basements or subway stations, it would remain dark come morning. Even if physicians could get to those who needed help, the lack of electricity would prevent the administering of most medical treatment, which requires adequate illumination of the wounded sites. In addition, much surgery relies on numerous electrically powered support devices. While most hospitals have back-up electrical generators in case the power should fail, most shelter medical stations would not.

Transistor radios and telephones are expected to provide communication links between the sheltered survivors and the civil authorities. But few shelters would have telephone jacks, and direct blast and fire damage to radio-transmitting and telephone-

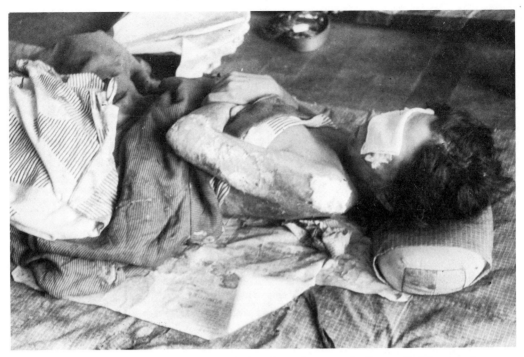

A woman waiting for help at a temporary relief station, where no effective medical treatment was available. At night, pathetic voices could be heard in the darkness, with nothing to light the rooms. Nagasaki, late August, 1945. (Eiichi Matsumoto, Hiroshima–Nagasaki Publishing Committee.)

relay stations might knock both systems out, even if the electromagnetic pulse did not. There might then be no way to inform physicians of the locations of those who needed them or of the whereabouts of medical supplies.

Such a basic consideration as clean water would become a major obstacle to providing adequate medical care. The failure of electric power for pumps and the rupture of connecting pipes would cause loss of water pressure, even if most of the water distribution system was intact. All surface water within the local fallout area (including reservoirs more than 200 miles downwind from the bomb) would be suspect and could not be used until tested for radioactivity. Water would have to be most severely restricted to drinking and food preparation. Cleanliness would be unattainable, and minimal medical standards impossible. Furthermore, all food not in airtight containers would be suspect. Who would know whether it was safe to eat an apple? A peeled apple? A piece of bread in an unsealed wrapper? With what water might one clean a chicken for cooking? People in shelters would soon run out of food, necessitating foraging trips out into the radioactive landscape. Who would go? The able-bodied, who might be needed more later? The expendable? Volunteers? It has been suggested that those over 40 should serve this purpose, as their life expectancies would be shorter than the latency periods before they contracted cancers (Gant and Chester, 1981).

Burial of the dead and disposal of wastes would be essential if disease were to be prevented and if significant psychological problems among the survivors were to be averted. A 1962 study published in the *New England Journal of Medicine* noted: "There is evidence that profound emotional disorders and somatic manifestations follow the sight and smell of decomposing bodies." To illustrate the magnitude of this task, it took eight weeks to bury 39,000 Japanese and Filipinos after the U.S. Army entered Manila in 1944, without the added problem of radiation exposure. Grave diggers, with few exceptions, experienced nausea, vomiting, and loss of appetite—symptoms that duplicate the early stages of radiation sickness. The circumstances unique to nuclear war would complicate prompt burial of the dead. Many bodies would be radioactive and would present a danger to those who came near them. Without accurate meters, it would be impossible to determine which were dangerous to approach.

Conclusion

When renal dialysis apparatus first became effective, the number of machines was far fewer than the number of patients with terminal kidney failure. To determine who would live and who would die, "God Committees" were established to make the life-and-death decisions. The ethical questions created by the selection of patients for renal dialysis imposed a heavy burden on the members of the God Committees, but compared to the responsibilities of physicians in the post-nuclear-attack period, such dilemmas seem relatively insignificant.

Should the individual doctor make his way toward the center of devastation, providing succor to the agonies of the wounded and dying as he goes? Or, mindful of his special knowledge and skills

Rescue workers using grappling hooks to fish bodies from Tokyo's Sumida River. Thousands drowned seeking refuge from the smoke and flames during the firestorm generated by the conventional bombs dropped on Tokyo. The nausea, vomiting, and loss of appetite that result from disposing of decomposing bodies would mimic the symptoms of acute radiation exposure and present diagnostic problems to health care workers. March 1945. (Koyo Ishikawa, Tokyo.)

and the need for them in the aftermath of the nuclear horror, should he seek shelter until the radiation hazard had diminished? Would the doctor's concern for his family's safety dominate his actions? Or would he consider his responsibility to his fellow citizens more important than his personal needs? As he proceeded either in the devastated city or along the corridors of some temporary medical facility, should he decide to treat the patient who may be more important socioeconomically, the one who might have special skills necessary for the recovery of society (if these facts were known), or the one who is less critically injured?

Would physicians be psychologically capable of practicing triage, even if it were possible to distinguish those who would benefit from medical treatment from those who would not? Would physicians have the emotional strength to treat the injured without being overwhelmed by their own personal tragedies—loss of their homes, families, and friends? Would physicians have the stamina to perform their specialized skills in the face of tens or hundreds of thousands of casualties, and severely limited supplies of medications, medical facilities, ancillary personnel, and even safe food and water? Would the future appear to hold enough hope to make any heroic efforts seem worthwhile?

References

Arms Control and Disarmament Agency. 1979. "Effects of Nuclear War." Washington, D.C.

Blocker, Virginia, and T. G. Blocker. 1949. "The Texas City Disaster: A Survey of 3000 Casualties." *American Journal of Surgery*, November: 756–771.

Camp, F. R. 1973. *Military Blood Banking*. Fort Knox, Kentucky.

"Case Study: Care of Burned Children." 1979. *American Pharmacy, 19:* 19–33.

Eiseman, B., and N. Bound. 1978. "Surgical Care of Nuclear Casualties." *Surgery, Gynecology and Obstetrics, 146:*877–883.

Gant, Kathy S., and Conrad V. Chester. 1981. "Minimizing Excess Radiogenic Cancer Deaths after a Nuclear Attack." *Health Physics, 41(3):* 455–463.

Hiatt, Howard H. 1980. Statement before the Subcommittee on Health and Scientific Research of the Committee on Human Resources, U.S. Senate, June 19.

Office of Technology Assessment. 1979. *The Effects of Nuclear War.* Washington, D.C.: U.S. Congress.

Orth, G. L. 1959. "Disaster and Disposal of the Dead." *Military Medicine, 124:* 505–510.

Rice, R. M. 1956. "Medicinal Supplies for Mass Casualties from Pharmaceutical Producer's Viewpoint." *Military Medicine, 118:*262. Cited in Saul Aronow et al., eds. 1963. *The Fallen Sky: Medical Consequences of Thermonuclear War.* New York: Hill and Wang, p. 29.

Sandler, S. L. 1977. "Blood Transfusions Therapy During the Yom Kippur War, October, 1973." *Military Medicine, 142:49–53.*

Schildt, Ever. 1967. *Nuclear Explosion Casualties.* Springfield, Ill.: Thomas.

Sidel, Victor, H. Jack Geiger, and Bernard Lown. 1962. "The Physician's Role in the Postattack Period." In "The Medical Consequences of Nuclear War." *New England Journal of Medicine, 266:1126–1155.*

Surgeoner, Douglas, M.D., Director, Northeast Regional Red Cross. Interview by Holmes Morton, James Anderson, and Andrew Levitt. January 1981.

15

Burn Injuries Among Survivors

John D. Constable, M.D.

The crash of a partially filled 30-passenger airplane on an island off the coast of Massachusetts required the mobilization of all the emergency medical facilities of Greater Boston, a major surgical center. Yet we are asked to contemplate the possibility of ten thousand, or a hundred thousand, or even a million severely traumatized victims of a military nuclear explosion. Adequate medical treatment for such survivors is an impossibility.

We can talk about how such injuries *should* be treated, but to transfer this knowledge to the practical possibilities of treating the number of victims that have been predicted is categorically out of the question.

The injuries caused by a massive nuclear detonation would come from the various effects of such an explosion. People within and around buildings would suffer extensive traumatic injury, from being blown out of the buildings and from damage by debris. The initial blast effect of the explosion is characteristically followed by powerful winds rising to as much as 180 miles per hour, which would cause a number of severe traumatic injuries.

Most of those who have been crushed, cut, or blasted, but not burned, and who have survived initial injury and reached medical facilities would, in most cases, be expected to require only one major surgical procedure. Although this might be very expensive in terms of time and material, including a great deal of blood and other support, the victims could then be expected to enjoy a relatively uncomplicated convalescence.

Such would not be the case for burns. Even though heat and light contain only some 35 percent of the total energy of a nuclear explosion, burns would consume a far higher percentage of the post-attack medical resources.

Let us consider the burn injuries that would result from a nuclear explosion and the treatment that would be required.

Patients suffering from anoxia, resulting from most of the atmospheric oxygen having been used up by the extensive fires (especially firestorms), would be rare. If the degree of thermal activity was sufficient to have caused anoxic damage, then usually there would be concomitant fatal incineration.

Both carbon monoxide poisoning and fire-induced anoxia must be distinguished from *pulmonary burns*, which remain one of the major therapeutic problems of thermal damage, one that is largely unsolved. This form of lung injury usually takes from 24 to 72 hours to develop and is not the result of direct thermal damage to the lung. If the heat around the patient's face is sufficient to actually destroy the trachea, bronchi, or lungs, there is almost invariably such devastating destruction of the face and other parts of the skin that the patient does not survive.

The generally accepted theory is that the damage to the lungs results from the chemical activity of noxious products of incomplete combustion. Consequently, this type of burn is characteristic of fires in closed spaces rather than the open spaces that would be more common with a major bomb. Among people confined to buildings, pulmonary burns would be a major lethal factor. In the Coconut Grove fire in Boston some 40 years ago, more than 400 people died, almost all without visible signs of burns. These deaths, which occurred mostly two, three, and four days after the fire, resulted from pulmonary damage now believed to have been from the fumes from the plastic in the artificial palm trees and furniture coverings.

203

Two kinds of direct thermal injury would occur from a nuclear explosion: one directly from the detonation, the other from the secondary fires following the ignition of available combustible material. These secondary fires would be of at least two sorts. One possibility is a firestorm. Much more certain is the development of a major conflagration, which would be essentially the sort of fire with which we are all too familiar, but enormously increased in scale. This fire would be associated with multiple smaller ones, starting from the breaking of gas mains, the failure of electrical pumps, the lack of water to put them out, and so on. The fires would be spasmodic over a very large area.

People would be exposed to the risks of thermal damage from the bomb itself and from its secondary fires. There is no essential difference in the nature of burns from these two etiologies. Burn damage to the skin results from a combination of the amount of heat and the time of exposure, these factors being very much modified by the presence or absence of clothing, the moisture content of the atmosphere, and other factors. An explosion results in an almost instantaneous exposure to a very high heat level, with damage occurring over an incredible distance; but the nature of the injury is not different from other forms of thermal burns. It simply means that there can be much more severe damage in a very short time if the heat to which one is exposed is very great.

All people seriously injured by a nuclear explosion who also have had a significant amount of radiation injury would be more difficult to treat. Some survivors would have received sufficient radiation to result in death within a matter of weeks or months from the radiation alone. But even with those who received smaller doses of radiation, the damage to the immune system and to blood element regeneration would result in the patient being more prone to invasive sepsis, in less satisfactory healing, and in an increased risk of death from a thermal injury that might otherwise not have been fatal.

Experimental studies have shown that a burn from which a normal animal can be expected to recover becomes lethal if the animal has been previously or concomitantly exposed to non-lethal radiation.

First-degree burns at their very worst are equivalent to a severe sunburn. They may result in some transient dehydration, certainly considerable pain, but under any emergency conditions these require essentially no treatment and must be considered of no particular medical consequence.

204

Second-degree or partial-thickness burns (the latter term is much to be preferred) are, from the point of view of the surgical problems, almost as severe an injury as are third-degree or full-thickness burns. A deep partial-thickness burn requires essentially the same amount of resuscitative effort, the same difficult nursing, the same elaborate dressings, and the same extensive care during the first three to four weeks. Although these injuries heal from the base and therefore no skin grafting is required, and the eventual problems of resurfacing the patient are a great deal simpler, the immediate problem of care is almost as great as with a full-thickness burn. The two groups should be combined from the point of view of trying to evaluate the early load on the medical system.

Estimating accurately the extent and number of burn survivors in a population exposed to a nuclear explosion is very difficult. The figure might vary by as much as a thousandfold, depending upon specific factors prevailing at the time of the explosion. Even a moderate degree of opacity in the air strikingly reduces the range of thermal damage. Other factors include the season, the time of day, and the extent to which the population had been warned. These conditions partly determine the amount of clothing being worn and whether people are outdoors or not.

For a one-megaton nuclear explosion, with 10-mile visibility, it has been estimated that third-degree or full-thickness burns might be expected within 5 miles; second-degree or partial-thickness burns within a 6-mile radius, and first-degree burns within 7 miles. If the atmosphere were sufficiently opaque to reduce visibility to 2 miles, then the second-degree zone would be

---

allograft   a graft taken from another person, not an identical twin.

anoxia   absence of oxygen.

sepsis   presence of various pus-forming and other pathogenic organisms, or their toxins, in the blood or tissues.

triage   classification of a large number of casualties into three groups: those who will survive without any medical help, those who will die no matter what treatment they receive, and the priority group of those who will survive only if they receive medical treatment.

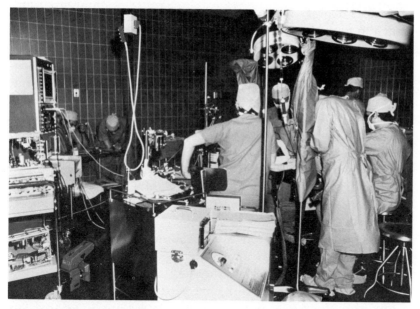

Burn treatment. Large numbers of highly trained personnel are required. Such personnel will not be available after a nuclear war. (Shriners Burns Institute, Boston.)

reduced from 6 miles to something under 3 and the others changed proportionately.

Unfortunately, burns are the form of trauma that characteristically demand the largest amount of medical assistance. No injury can be counted on to use up more hospital facilities than a severe burn. Triage would be very difficult, and a great many patients treated for extended periods might still eventually die from their injuries.

The burn literature has been filled over the last ten years with reports of progress in salvaging the severely burned. Many new methods of infection control have come into use, including various surface antiseptic agents and topical antibiotics. The surface control of infection has prevented the conversion of partial-thickness to full-thickness burns by sepsis and has strikingly improved overall results in burn salvage. There also has been much effort to control systemic infection, both by the use of antibiotics and by elaborate isolation techniques. "Life islands," in which patients are isolated in a plastic enclosure, and laminar flow units, in which the air is regularly replenished and replaced so that

206

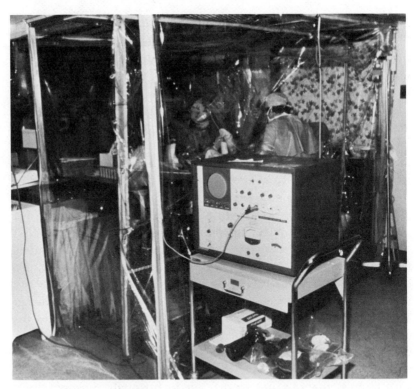

Isolating the burn victim. A plastic tent provides a sterile environment for the treatment of a severe burn. Pathogenic organisms are screened out both by the tent and by a constantly recirculating air flow filtration system inside. (Shriners Burns Institute, Boston.)

bacteria are swept away, are recent innovations. All of these methods have helped reduce death from infection.

Another recent development is the early surgical excision of burns. Although it is usually not safe to excise more than one-fifth of the patient's body surface at one time, surgery may be carried out on the first or second day after the burn, and with maximum support again on the fourth, and so on, ending with as much as 80 to 90 percent of the skin being excised. Massive excision has been combined with immunosuppression to allow for the use of typed allografts taken from living donors or cadavers. Some dramatic results are possible with these methods, although they are still cosmetically relatively grotesque.

It is absolutely essential to recognize that any really severe burn may require as many as 30 to 50 operations, both immediate and delayed, and months and months of hospitalization. This care imposes immense strains on the medical facilities available. With the newer and more dramatic methods, there is at least the possibility, if sufficient material and personnel are poured in, of salvaging burns in the 85 to 90 percent range.

Triage is much more difficult when the physician is faced with an enormous group of patients sustaining 20 to 90 percent burns who might survive if treated. (Except for burns of the hands and face, I exclude burns affecting under 20 percent of surface because most of these can be treated relatively easily.) What is involved in treating large numbers of severe burns?

Some years ago the Shriners of North America, who had for years donated large sums to look after orthopedically crippled children, became interested in building specialized burn hospitals for children. Their plan was to start with three burn units and then to expand, possibly adding another 15 or so to match the number of orthopedic hospitals they were already maintaining. These initial three units were built in Boston, Galveston, and Cincinnati. In the 15 years since these three 30-bed hospitals were built, it has not been practical to build even one other unit, because the three burn units, with a total of 90 beds, consume a budget similar to that of nineteen orthopedic hospitals, most of which are of comparable size.

The cost of running a single 30-bed hospital, where half of the beds are reconstructive and where there would rarely be more than ten acute burn cases at one time, is in the neighborhood of $4 million per year. The United States has approximately 1000 to 2000 so-called burn beds in specialized institutions. Each burn patient requires specialized individual nursing for quite a long time. At most, one nurse can look after two patients.

Severe burn cases require not just one major operation but may need general anesthesia every other day and regular trips to the operating room for weeks or even months. Elaborate dressings and the application of antibiotics or at least antiseptic agents are necessary. The patients require large amounts of blood, albumin, and other human blood derivatives. They may need enormous areas of allografts, but after a nuclear war, obtaining sufficient quantities of these from cadavers may be difficult, because many of the dead would be in highly radioactive areas or be contaminated radioactively themselves.

Whereas most traumatic lesions are treated definitively imme-

208

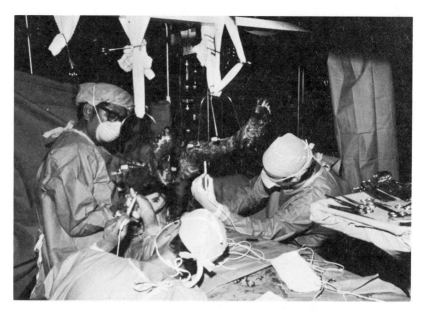

A child with severe leg burns requiring extensive surgical treatment in a highly specialized hospital. Severe burn cases may need regular trips to the operating room for weeks or months. (Shriners Burns Institute, Boston.)

diately, and the victims either recover or die, burns are peculiar. The burn patient is not so ill during the first 12 to 24 hours. I have seen a number of older patients with 40 to 50 percent full-thickness, clearly fatal burns who, for the first 12 to 24 hours after their injury, appeared in reasonably good condition. They were capable of consulting their lawyers and doing whatever needed to be done. After this initial period the patient becomes sicker and sicker, and this critical hovering between survival and death may go on for weeks or months.

Once a burn has been initially resurfaced, it may need months or years of reconstruction. And even with all of this, anyone who is discriminating or humane would recognize that the end results are indeed pathetically poor. It is very difficult to estimate the cost of such cases in dollars because, to the best of my knowledge, no health program or insurance pays adequately for burn care. Blue Cross/Blue Shield and similar insurance programs admit that they cannot afford to pay the true cost. Nonetheless, it is reasonable to put the cost at anywhere from $200,000 to $400,000 for a severe surviving burn case.

Treatment of a Severe Burn

20-year-old man

85% of body surface: full-thickness burns (from a gasoline explosion)

33 hospital days (died on 33rd day)

501 transfusions:

fresh frozen plasma	281 units
fresh frozen red blood cells	147 units
platelets	37 units
albumin	36 units

6 operations

medical personnel hours	4900
approximate daily cost	$3500

From John F. Burke, M.D., Chief of Trauma Services, Massachusetts General Hospital, July 1980.

Even though there are 30-bed burn units, such as the Shriners or those at large general hospitals, they can handle only two or three fresh severe burns at once. If a large group of such burns occurred in a major accident, they would have to be distributed for effective treatment.

Major burn disasters of recent years—the Coconut Grove and Hartford Circus fires—and various plane crashes have resulted in very few survivors of major burns. Initially following a nuclear attack, there would be thousands, or even tens of thousands, severely burned immediate survivors. Even the most conservative calculation of thermal injuries resulting from an isolated one-megaton or "minimal" nuclear explosion, with hypothetical preservation of all U.S. medical facilities and the availability of immediate and perfect triage and transportation, shows that what we consider to be one of the most lavish and well-developed medical facilities in the world would be completely overwhelmed. It is impossible to imagine the chaos that would result from a larger explosion in which the hospitals themselves were partially destroyed and where there was no possibility of significant triage or inter-center transportation. The medical facilities of the nation would choke totally on even a fraction of the burn casualties alone.

16

Survivors of Nuclear War: Infection and the Spread of Disease

Herbert L. Abrams, M.D.

The devastation and chaos that would follow an all-out nuclear war and the potential for regression to a social structure unknown to twentieth century industrialized society have been well emphasized. But the nature of the medical problems that would confront survivors has not been widely conveyed. The effects of burn, blast, and radiation have dominated discussions of the post-attack period. In the intermediate term, however, infection and the spread of communicable disease would represent the most important threat to survivors.

In depicting a massive nuclear exchange, we will assume that the United States has undergone a 6500-megaton attack, the so-called CRP-2B model used by the U.S. Federal Emergency Management Agency in civil defense planning. In terms of yield, it

This manuscript was written while the author was a Henry J. Kaiser Senior Fellow at the Center for Advanced Study in the Behavioral Sciences, Stanford, California (1980–1981). The author would like to express his gratitude to the Center and to the Henry J. Kaiser Family Foundation for their support during this period. Much of the material was included in an article in the *New England Journal of Medicine* (*305*: 1226–1232, 1981) and as a chapter in *The Final Epidemic: Physicians and Scientists on Nuclear War* (Chicago: Bulletin of the Atomic Scientists and University of Chicago Press, 1981). The figures on pages 217, 221, and 230 are not from the original article. The author also acknowledges the assistance of William E. Von Kaenel.

Table 1. Casualties from a 6500-megaton attack on the United States.

Cause	Fatalities (millions)	Injured or Affected (millions)
Trauma, flash burns	86	34
Trapped by debris	5	2
Secondary fires	3	1
Lack of water	1	4
Inadequate shelter ventilation	1	5
Fallout radiation	38	23
Net casualties	134	33

Source: U.S. Federal Emergency Management Agency, 1979, p. 3. This publication gives the pre-attack population as 237 million. Since our present population is about 225 million, the figures were all adjusted accordingly, so as not to differ from other casualty estimates used (by FEMA) that assume about 225 million. Those "affected" by fires, water, and lack of ventilation are said to be "forced out [of shelter] by . . . ." This implies they might have been killed by radiation as much as by actual burns, dehydration, or suffocation.

represents 524,000 Hiroshima bombs. The targets of attack, in order of priority, would include:

- military installations
- military-supporting industrial, transport, and logistic facilities
- other basic industries and facilities that contribute to the maintenance of the economy
- population concentrations of 50,000 or greater

Some 4000 megatons would be detonated on urban areas and population centers. Moments after the attack, 86 million people— nearly 40 percent of the population—would be dead (Table 1). An additional 34 million—27 percent of the survivors—would be severely injured. Forty-eight million additional fatalities are anticipated during the shelter period, for a total of 134 million deaths. Many of the millions of surviving injured would experience moderate to high radiation doses and would have residual blast and burn injuries.

Table 2 lists the medical problems resulting from this 6500-megaton attack. The periods under consideration are as follows:

Table 2. Medical problems during the attack and in the post-attack period.

Medical Problem (in approximate order of time in which inflicted)	First Hour	Shelter Period		Survival Period	Long-Term Effects	
		First Day	First 0–4 Weeks		Recovery Period	Future Generations
Flash burns	+					
Trauma	+					
Flame burns and smoke inhalation	+	+				
Acute radiation	+					
Fallout radiation	+	+	+	+		
Suffocation and heat prostration		+	+			
General lack of medical care		+	+	+		
Dehydration			+			
Communicable diseases			+	+		
Exposure and hardship			+	+		
Malnutrition			+	+		
Cancer					+	
Genetic effects						+

Immediate Effects. During the barrage period, the explosions almost instantaneously would inflict millions of lethal and non-lethal blast, thermal, and immediate radiation injuries on those caught in and around the blast areas.

Shelter Period. From the time of the attack to days or weeks later, those surviving the initial explosion would attempt to sustain themselves in fallout shelters, amid intense radiation, fires, and deprivation.

Post-Shelter Survival. Fallout would reach a level "acceptable" for emergencies after variable time periods. The problems of attaining food, finding shelter, and recovering from acute injury would have to be confronted. The injured would need to be nursed, the dead buried, debris cleared, the harvest reaped, and the next har-

213

> I refuse to accept the cynical notion that nation after nation must spiral down a militaristic stairway into the Hell of nuclear destruction.

Martin Luther King, Jr., 1964

vest sown. In a hazardous environment, survival would be the only meaningful goal.

Long-Term Effects. Survival would be accomplished and some kind of recovery initiated. Some societal structure would emerge, food supplies secured, shelter obtained, and communities established. A primitive social organism would be strained by intense competition for food supplies. During the early years, the first cases of radiation-induced leukemia would appear; later the solid cancers would develop in the lungs, thyroid, breast, and colon.

The problems of infection and communicable disease would be severe during the shelter period and more particularly during post-shelter survival. They require careful consideration, not only because of human cost, but also because of their impact on the recovery process. What, then, is the nature of the threat: why is the likelihood of infection so much greater in the post-attack world?

Increased Risk and Severity of Infection

Most survivors would experience increased susceptibility to infection because of both the pervasive direct effects of nuclear weapons and the subsequent pressures and hardships confronted. Several factors would be of special importance.

Radiation

Radiation affects the immune system in a number of different ways, not least of which is its capacity to injure the bone marrow and the lymph nodes. Hematologic changes may occur with doses as low as 50 rems: decreased antibody response, decreased effectiveness of cellular defense mechanisms, decreased effectiveness of immunizing agents, and increased susceptibility to some toxins. Consequently, vaccination would be less effective.

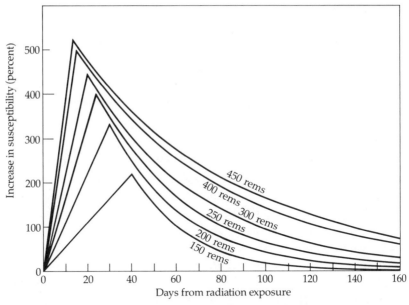

Effect of radiation exposure on disease susceptibility over time. (From Voors and Harris, 1970.)

Radiation also has a major effect on the lining or mucous membrane of the intestine. The ulcerated lining provides a portal of entry for intestinal bacteria into the bloodstream, with bloodstream infection as a certain sequel. These organisms are frequently difficult to control with antibiotics.

Twenty-three million survivors of a large-scale nuclear war would suffer from radiation sickness, evidence that they received a mean dose of 200 rems or more. But the number of cases with doses between 100 and 200 rems would probably equal that number, so that 50 percent of the surviving population might have lowered resistance to disease from radiation exposure alone, extending over a period of many months (see the figure above).

Trauma and Burn Casualties

Among the millions suffering from trauma and/or burns, over a third also would suffer from radiation sickness. Aside from the risk of infection related to the open wounds, weakness and incapacitation could be expected to increase general vulnerability. There is also a known synergy between burns and radiation that profoundly increases the mortality rate.

215

Malnutrition and Starvation

During the shelter period, the availability of food would vary. Adults can maintain health for several weeks with only minimal food intake, but in some locales, fallout might prevent emergence from shelters for longer periods of time so that nutritional health would deteriorate. Infants and young children, in particular, might experience severe malnutrition due to insufficient or inappropriate foods during an extended shelter stay.

During the post-shelter phase, three crucial periods can be defined: shelter emergence until the first harvest, first harvest until the next harvest, and all subsequent seasons. The first two periods depend a great deal on the time of year the attack occurs, and the third depends on post-attack recovery and environmental conditions. Upon shelter emergence, most of the food stores would have been destroyed in urban areas; remaining supplies most likely would be consumed during or soon after the shelter period.

The essential supply of available food would be the grain stored in small towns and rural areas. The lives of millions of survivors would depend upon this supply until the next harvest was avail-

---

beta-hemolytic *Streptococcus* a type of bacterium that most frequently causes acute infections, such as tonsillitis, but can result in serious complications involving the heart (acute rheumatic fever) and kidneys (acute glomerulonephritis).

bubonic plague an acute infectious disease, transmitted to humans by infected rats and other rodents, characterized by lymph node enlargement and high mortality.

endemic disease infectious disease prevalent in a particular locality.

pneumonic plague a highly infectious and almost always fatal form of plague involving the lungs and transmitted from human to human without a rat vector.

rem 450 rems of total body irradiation is considered the dose that will kill approximately 50 percent of the healthy adults exposed.

World War II: the face of hunger, during the siege of Leningrad. A starving man holding his daily ration of bread. The bread was often made from ingredients such as moldy flour, cellulose, and cotton seed. (Sovfoto.)

able. The amount of stored grain varies considerably during the course of the year. Supply for the surviving population might vary from 200 to 500 days, with great dependence on the next harvest.

But food piled high in silos in remote regions would do little good for a hungry populace. Grain would have to be obtained, transported, and distributed to the survivors where they were located. Grain transportation would be the most important survival activity in the immediate post-shelter period. This problem would be made much more difficult by the negative correlation that exists in the United States between population and grain density (see the figures on the facing page).

Furthermore, assuming that trucks and highways would be usable, an essential commodity for the transportation of grain is fuel. It is estimated that as much as 99 percent of U.S. refining capacity could be destroyed during the 6500-megaton attack.

Exposure and Hardship

Widespread destruction of urban housing would occur, with major damage to rural housing as well. Heating fuel might be unavailable. General hardship, with exposure, poor nutrition, and exhaustion increased by enormous physical demands, would follow and would promote great vulnerability to infection.

Lowered Natural Resistance to Disease

Surviving Americans would for the first time experience the underdeveloped world as their natural habitat. Unlike the population of impoverished lands, however, Americans do not have the high natural immunity to a host of dangerous diseases that allows many in the Third World to survive. The omnipresence of antibiotics has altered the normal production of antibodies to infectious agents among the developed nations. Because of the destruction of the pharmaceutical industry, as well as post-attack disorganization and chaos, antibiotics would be in short supply for countries that have depended on them.

Successful campaigns to eliminate lethal epidemics, such as cholera and typhoid fever, have been accompanied by a failure to develop resistance to these diseases. Reintroduction of such "ex-

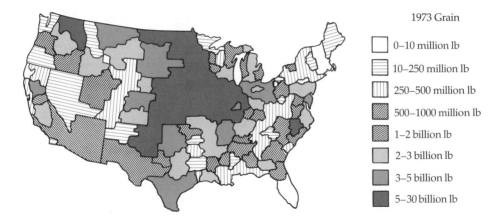

U.S. grain density. (From Haaland, Chester, and Wigner, 1976, p. 149.)

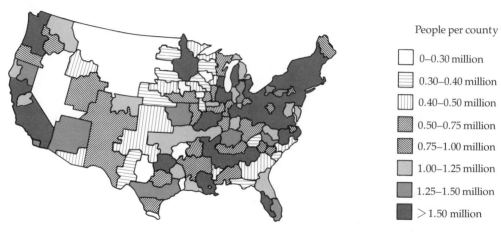

U.S. relocated-population density. (From Haaland, Chester, and Wigner, 1976, p. 146.)

otic" diseases might find the population incapable of handling them, as were American Indians when exposed to the diseases of Europeans. Measles, whooping cough, and diphtheria might run rampant in un-immunized infants, and beta-hemolytic streptococcal infections would be widespread.

Factors Increasing the Spread of Disease

Shelter Conditions

Large public shelters probably would operate under severe limitations, with thousands packed into inadequate areas. Under these circumstances, hepatitis and respiratory and gastrointestinal infections might spread rapidly. Most shelters would lack adequate ventilation. When ventilation systems were present, the blowers and fans could easily be rendered inoperable by an overpressure of 1 psi and the systems blocked by electric power failure. Heat and humidity in the shelter would then increase, and the absence of a continuous flow of fresh air would encourage the spread of infective microorganisms.

The length of time that fallout radiation would enforce basement and underground occupancy would be an important determinant of disease spread. This period might be several months. Those in highly radioactive cities would have much longer stays and might have significant exposures even in shelters. Even after it became permissible to work outside, it might still be necessary to eat, sleep, and rest in fallout shelters.

Sanitation

The barriers to communicable disease spread accomplished today by a sanitary water supply, properly prepared and refrigerated food, sewage treatment, and waste disposal would be seriously compromised in the post-attack environment.

Impure water, contaminated food, and general unsanitary conditions would spread a host of intestinal diseases not yet experienced by most Americans. These include infectious hepatitis, *E. coli*, *Salmonella*, *Shigella*, amebic dysentery, and possibly typhoid and paratyphoid.

220

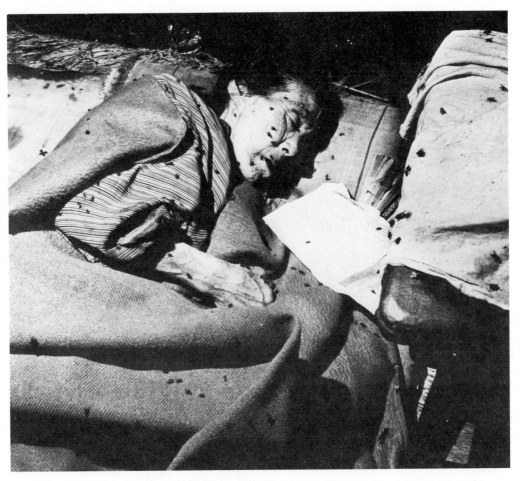

An old woman and flies. Immediately after the bombing, there was an increase in the number of flies, and maggots got into the wounds of many victims. Hiroshima, September 1945. (U.S. Navy, Hiroshima; Hiroshima–Nagasaki Publishing Committee.)

Insects

Insects are generally more resistant to radiation than humans. This resistance and the existence of corpses, waste, lack of sewage treatment, depletion of birds, and destruction of insecticide stocks and production would engender a huge increase in insect growth. The absence of control of insect growth, combined with failure to provide adequate sanitation, might sharply limit the capacity to control such diseases as typhus and dengue fever.

Corpses

The health problem created by millions of corpses post-attack would represent a serious disease threat. In many areas radiation levels would be so high that corpses would remain untouched for weeks on end. With transportation destroyed, weakened survivors, and a multiplicity of post-shelter reconstruction tasks to be performed, corpse disposal would be enormously complicated. To bury the dead, after a 6500-megaton attack, an area 5.7 times as large as the city of Seattle would be required for the cemetery.

Factors Limiting the Response to Infection

Government Organization

The United States has developed an extraordinary ability to take effective countermeasures against communicable diseases. Should an outbreak occur, a public health network is informed, the rest of the country alerted, and appropriate steps taken. In 1947 a man infected with smallpox mingled with New York City crowds for several days. More than 6,350,000 persons were immediately vaccinated; as a result, only 12 additional cases appeared. More recently, the unfortunate swine flu episode illustrated how the hint of an epidemic could bring enormous medical resources to bear upon the threat.

Coherent efforts to control and limit the spread of disease require surviving government, organized geographic units, communication networks, and a favorable enough survival situation so that physicians and health officials can perform their tasks.

All of these conditions are speculative in the post-attack world. Most radio contact would be eliminated by nuclear weapons effects. Treating the wounded would require the full attention of available medical resources. For a sustained period, surviving officials might have to remain in shelters, with the acquisition and supplying of food and water their primary concern. The huge number of injured, the tenuous food situation, massive industrial destruction, enormous debris removal and body disposal tasks, and disparity between "food-rich" and "food-poor" regions would undermine interregional cooperation seriously.

222

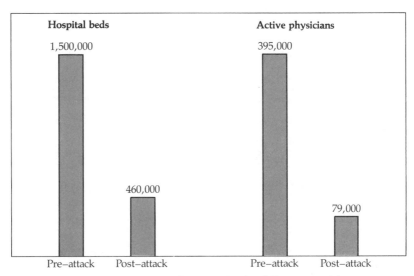

U.S. medical resources after a nuclear war. (Federal Emergency Management Agency, 1980, p. 63.)

Disease Detection, Diagnosis, and Treatment

Health countermeasures against potential epidemics depend upon the availability of resources and the involvement of physicians. Many hospital beds would be destroyed, and casualties among physicians and other health personnel would approximate 80 percent (see the figure above). This percentage is higher than the casualty percentage for the population as a whole (73 percent) because physicians are disproportionately represented in large cities.

If we assume that the average shelter contains 100 people, and that no more than one physician would be in any shelter, then, using U.S. government estimates for surviving, uninjured physicians, only one in twelve shelters would have a functioning physician. If a serious epidemic occurs, it would affect physicians as well. More physicians would become incapacitated as the number of sick increased, further raising the injured to physician ratio.

Laboratories are essential for dealing with communicable disease, but they would be highly vulnerable to the effects of an attack.

Table 3. Principal communicable disease countermeasures by mode of transmission.

Countermeasure	Mode of Transmission		
	Intestinal	Human-to-Human	Vector-borne
Antibiotics	+	+	+
Excreta disposal	+	−	−
Food hygiene	+	−	−
Immunization	+*	+	−
Potable water	+	−	−
Public information	+	+	+
Vector control	+	−	+

*Typhoid fever only.
Source: Johnson et al., 1978.

Thus the prompt detection and diagnosis of communicable diseases essential to the identification of medical resource needs would be impeded severely during a crucial period.

Inadequacy of Countermeasures

Many of the countermeasures required for intestinal and vector-borne diseases would be unavailable in the disorganization of the post-attack period. Their control involves assuring supplies of pure water and uncontaminated food, disposal of sewage and waste, and removal of breeding areas for insects and rodents (Table 3). Both antibiotics and immunization would play an essential role in stemming epidemics. But how effective would they be?

Antibiotics. Antibiotics are ineffective in combating viral disease and cannot limit the spread of such infections as smallpox, viral gastroenteritis, and influenza. Several dangerous bacterial diseases such as diphtheria and tetanus respond poorly to antibiotics. Furthermore, the demand for antibiotics would be large. If laboratory tests were not available, they would be prescribed for all undiagnosed ailments. Most stores of antibiotics in urban centers would be destroyed, and those still intact might be inaccessible for days or weeks because of intense fallout radiation. The pharmaceutical industry would be virtually eliminated in a massive attack.

Immunization. For several hazardous diseases, such as tetanus, poliomyelitis, measles, whooping cough, and typhus, immunization is the only effective direct means of control. In post-attack conditions, however, the effectiveness of vaccination programs would be diminished by the impact of radiation on the immune system. Millions who had substantial radiation doses and therefore needed immunization most of all would benefit least.

Potential Pathogens in the Post-Attack World

Studies performed in the late 1960's identified 23 diseases that might be significant in the post-attack environment, 18 of which are listed in Table 4. Many of these are encountered in endemic form throughout the country. Among them, potential epidemic sources may be divided into two categories. The first includes the classic epidemic diseases, fortunately of low incidence; the second, diseases of heightened incidence but low mortality (Table 5). Respiratory diseases, including viral pneumonias, influenza, pneumococcal and streptococcal infections, and tuberculosis infections would affect particularly those living in crowded fallout shelters, with an augmented impact on the young and the old. The diarrheal diseases caused by the bacteria *Salmonella*, *Shigella*, and *Campylobacter* and viral gastroenteritis would be widely prevalent. Although their mortality rate is usually low, in the presence of radiation injury to the gastrointestinal tract it would be increased substantially. Furthermore, these diseases, as well as infectious hepatitis, spread rapidly in the absence of adequate sewage disposal, pasteurized milk, or appropriate sanitary precautions. The group of diseases endemic to rural areas, and thus a danger to evacuated populations, includes rabies, plague, and tetanus. Other diseases such as cholera or influenza might spread rapidly in devastated areas.

A more detailed view of two among many diseases that are generally considered well controlled in Western society will indi-

Table 4. *Infectious diseases in the post-attack period.*

Communicable Disease	Reported Cases in the United States in 1979
Amebiasis	4107
Diphtheria	59
Encephalitis, arthropod-borne, viral	1266
Food poisoning (botulism)	45
Food poisoning (salmonellosis)	33,138
Hepatitis, A	30,407
Influenza	0.3*
Measles	13,598
Meningococcal meningitis	2724
Plague	13
Pneumonia	19.7*
Rabies	4
Shigellosis	20,135
Smallpox	0
Tuberculosis	27,669
Typhoid fever	528
Typhus	69
Whooping cough	1623

*Death rate per 100,000 in 1979.
Source: Center for Disease Control, 1980, p. 3.

cate the roots of the concern for the role for communicable disease in the transformed post-nuclear-war world.

Tuberculosis

Tuberculosis, the Great White Plague of the nineteenth century, was a lethal infection for large segments of the population. Death rates as high as 550 per 100,000 were reported in New York City. If the annual U.S. death rate of 184.7 per 100,000 from tuberculo-

Table 5. Potential epidemic diseases.

Group 1: Epidemic Diseases of Low Incidence	Group 2: Serious Existing Diseases
Cholera	Diarrhea
Malaria	Diphtheria
Plague	Hepatitis
Shigellosis	Influenza
Smallpox	Meningitis
Typhoid fever	Pneumonia
Typhus	Tuberculosis
Yellow fever	Whooping cough

sis during the period 1900–1904 characterized our present population of 225 million, all the U.S. deaths from World Wars I and II, Korea, and Vietnam would be surpassed in one year and 10 days.

Should this concern us for the post-attack period, when we know that the mortality rate of tuberculosis has fallen below 1/200 of the 1900–1904 figures? In 1978 there were only 2830 deaths, and 28,521 new active cases in the United States. The percentage of the population who are positive reactors has also dropped dramatically to 4 to 8 percent of all tested.

But the bulk of this decline was achieved without the aid of drug therapy. By 1944, when the modern era of effective antituberculosis drug treatment began, the mortality rate had dropped to 43.4 per 100,000, less than 24 percent of its 1900–1904 rate. This change was largely attributable to improved socioeconomic circumstances, particularly since the incidence and mortality of tuberculosis rose and fell frequently in the past with altered societal conditions, especially in times of war (see the figure on p. 228).

During World War I, mortality increased 218 percent in Warsaw, reaching a rate of 974 per 100,000 in 1917. The highest incidence in all European cities in World War I was in Belgrade; the rate in 1917 reached 1483 per 100,000. During World War II the death rate rose 268 percent in Berlin, 222 percent in Warsaw, and 134 percent in Vienna. An analysis of 2267 chest roentgenograms at the Dachau concentration camp at the time of liberation

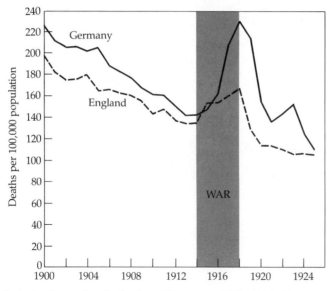

Tuberculosis death rate for England and Germany, 1900–1924. (From Mitchell, 1967.)

showed that 28 percent had evidence of tuberculosis, of which nearly 40 percent were "far advanced."

In the aftermath of nuclear war, all the factors that increase susceptibility to and spread of infection in general would be particularly applicable to tuberculosis. The destruction of housing, lack of fuel, shortages of food, medicine, and clothing, and a sustained period of labor and struggle would create precisely the setting in which tuberculosis has flourished in the past.

Nutritional status traditionally has been associated with an increased incidence and mortality from tuberculosis. More than 20 different studies have shown the relationship between food quality and tuberculosis, most striking during wartime. Animal protein is particularly important (see the figure opposite).

Fewer Americans have been exposed to tuberculosis today than ever before. Numerous examples of catastrophic epidemics among largely unexposed populations have been reported. South African troops in World War I had a mortality rate of 1745 per 100,000 from tuberculosis, while British troops had a rate of only 11 per 100,000. When Saskatchewan Indians were removed from nomadic to reservation life around 1880, their death rate from tuberculosis reached 9000 per 100,000.

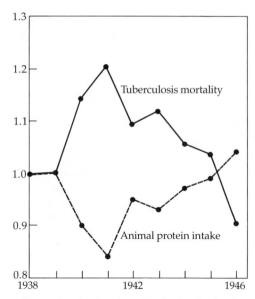

Tuberculosis mortality and animal protein intake for England and Wales, 1938–1946. (Adapted from Mitchell, 1967.)

In the United States from 1804 to 1808, eight of the top ten causes of death were infections (tuberculosis was number 1), accounting for 1526 deaths annually per 100,000 of population. From 1900 to 1904, six of the top ten causes of death were infections (tuberculosis was number 2), with 964 deaths per 100,000. By 1945–1949, only four of the top ten were infections (tuberculosis was number 7), accounting for 154 deaths per 100,000, just 10 percent of the deaths resulting from infection 140 years before.

Plague

Plague has been a known source of epidemics for the past 3500 years. In the twentieth century alone, more than 12 million deaths have been attributed to it.

Plague is endemic among wild rats in the 11 westernmost states. More than 30 types of wild rodents and rabbits have been found infected. Cats and dogs can be infected experimentally, naturally, or when they ingest infected rodents. Human contact with wild rodents is almost exclusively the source of plague cases in the United States, which accounted for 13 cases during 1981, 4 of which were fatal.

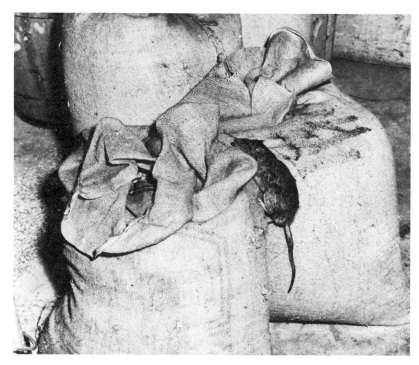

Rats in bags of grain. Plague is endemic among wild rats in the 11 westernmost states. (United Press International.)

A nuclear attack would create almost ideal conditions for breaching the "thin protective wall" against plague. Large areas of the western United States now relatively devoid of inhabitants might receive an influx of refugees from threatened or devastated urban areas. Relocation plans call for enormous increases in the population of many remote regions. Humboldt County, California, for example, would experience a fivefold or more increase in population. Millions of urban refugees, unable to obtain shelter in existing dwellings, would build earth-covered "expedient" shelters in undeveloped areas. Such shelters might provide good fallout protection, but they would create ideal conditions for transmission of plague from rodents.

Rodents are relatively resistant to plague, developing chronic infections, which function as a reservoir for the disease. Radiation would increase their susceptibility, as well as that of humans. High mortality among wild rodents then would help spread the disease to nearby humans: as the rodents die, the more resistant fleas would leave them and search for other hosts.

Once radiation had subsided in leveled cities, many survivors would head back hoping to reclaim whatever possessions might be found and to search for family members and friends. Over 90 percent of housing would be destroyed in a nuclear attack; there would be crowding in the remaining buildings. Conditions in the damaged cities would be favorable for the spread and propagation of plague. The rat population would increase because harborage and food for rats would be available.

A major danger comes not only from bubonic plague transmitted by domestic rats, but from human-to-human pneumonic plague. Under post-attack conditions, radiation and stress would raise the conversion rate of bubonic to pneumonic plague to 25 percent.

Quantitative Estimates of Infection in the Post-Attack Period

In the aggregate, deaths from communicable diseases among the survivors might approach 20 to 25 percent. Estimates of both the incidence and the mortality of infection in the post-attack world would vary widely for different diseases (see the figure on p. 232).

A computer simulation of the effects of a single nuclear explosion 9 miles south of New Orleans calculated that in the absence of medical countermeasures, 35 percent of the survivors would die from infectious diseases in the first year post-attack. This expected high mortality rate from communicable diseases would be ten times more than the normal death rate from noncommunicable diseases such as heart disease and diabetes, which would claim only 2.5–3 percent of the survivors. Equally, it would far surpass the cancer mortality, estimated at a few percent or less.

Conclusion

Numerous factors point to an increased risk of serious epidemics in the post-attack environment. These include the effects of irradiation, malnutrition, and exposure on the susceptibility to infection. Furthermore, unsanitary conditions, lengthy shelter stays,

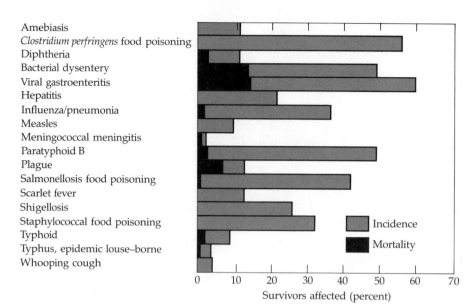

Disease incidence and mortality among U.S. populations exposed to nuclear weapons effects (with no medical countermeasures).

and insect population growth would facilitate the transmittal of disease. Depleted antibiotic stocks, physician shortages, laboratory destruction, and societal disorganization would render countermeasures largely ineffectual.

Although it is impossible to determine the extent of such "catastrophic" epidemics, it is certain that infection would pose a substantial threat to the health and recovery for all those injured by blast, heat, and radiation, and that the resources to grapple with this threat would be totally inadequate.

References

Abrams, H. L., and W. E. Von Kaenel. 1981. "Medical Problems of Survivors of Nuclear War: Infection and the Spread of Communicable Disease." *New England Journal of Medicine, 305:* 1226–1232.

Center for Disease Control. 1980. "Annual Summary." Morbidity and Mortality Weekly Report 28 (September).

Glasstone, S., and P. H. Dolan, eds. 1977. *The Effects of Nuclear Weapons,* 3rd ed. Washington, D.C.: U.S. Department of Defense and U.S. Department of Energy.

Haaland, C. M., C. V. Chester, and E. P. Wigner. 1976. *Survival of the Relocated Population of the U.S. After a Nuclear Attack.* Springfield, Va.: National Technical Information Service, 20–1. (Defense Civil Preparedness Agency report no. ORNL-5041.)

Johnson, D. R., M. L. Laney, R. L. Chessin, and D. G. Warren. 1978. *Study of Crisis Administration of Hospital Patients; and Study of Management of Medical Problems Resulting from Population Relocation.* Research Triangle Institute, RTI/1532/00-04F (September). Washington, D.C.: Defense Civil Preparedness Agency.

Mitchell, H. H. 1966. "Plague in the United States: An Assessment of Its Significance as a Problem Following a Thermonuclear War." Santa Monica, Calif.: Rand Corporation. (U.S. Atomic Energy Commission report no. RM-4868-TAB.)

Mitchell, H. H. 1967. "The Problem of Tuberculosis in the Post-Attack Environment." Santa Monica, Calif.: Rand Corporation, RM-5362 (June).

Sullivan, R. J., K. Guthe, W. H. Thoms, and F. L. Adelman. 1979. *Survival During the First Year After a Nuclear Attack.* Washington, D.C.: Defense Civil Preparedness Agency, 140. (Federal Emergency Management Agency report no. SPC-488.)

U. S. Federal Emergency Management Agency. 1979. *Analysis of Packages Comprising Program D-Prime.* Washington, D.C.

U. S. Federal Emergency Management Agency. 1980. "Material for the Record." U.S. Congress, Senatorial Hearing on Short- and Long-Term Health Effects on the Surviving Population of a Nuclear War, 96th Congress, 2nd Session, June 19.

Voors, A. W., and B. S. H. Harris. 1970. *Postattack Communicable Respiratory Diseases.* Research Triangle Institute, R-OU-493 (November), pp. 7–25. Sponsored by the Office of Civil Defense.

17

The Consequences of Radiation Exposure Following a Nuclear Explosion

Angelina K. Guskova, M.D.

The threat of nuclear war makes it a necessity that scientists all over the world carefully and responsibly analyze the scope of the impending danger. For 35 years, since the moment nuclear weapons were used in Hiroshima and Nagasaki, the effects of radiation exposure have been continually reassessed and made more precise. Data have been collected, for example, on cases of acute radiation sickness from accidental exposure. These cases have illustrated the relationship between the dose of ionizing radiation and clinical manifestations. Together with the findings from Hiroshima and Nagasaki, they allow predictions of the consequences of radiation exposure for a modern nuclear explosion.

Let us consider a one-megaton nuclear explosion. Up to one-third of the population of the targeted city would be killed directly, most dying during the first six days. Of the survivors, one-third to two-thirds would need medical care. The difficulties in delivering the necessary medical assistance, aside from the severity, numbers, and variety of medical problems caused by the explosion, would be enormously complicated by the deaths of and injuries to medical personnel, by the destruction of the majority of medical institutions, and by the abrupt diminution of food supplies and transportation.

234

Mother and child waiting to receive medical treatment. She sits in seeming oblivion, giving her breast to her baby. Nagasaki, between 2 and 3 PM on August 10, 1945, 3.6 kilometers (2.2 miles) from the hypocenter. "Straw mats are laid beneath trees in front of the railway station, where the injured are lying. They are brought here by hand carts. Moaning and sobbing can be heard constantly. One man stiffens and dies as he falls." (Yosuke Yamahata, Hiroshima–Nagasaki Publishing Committee.)

Treatment for isolated cases of acute radiation sickness with
bone marrow damage can be highly effective. However, the procedures and facilities required are highly sophisticated and would
not be available for the victims of a nuclear explosion. Proposals
that rely on the simplification of medical care in the post-nuclear-
attack period, with first aid provided by the general public or by
trained auxiliary personnel, do not take into account the extreme
complexity of treatment for the survivors of a nuclear war.

Acute Radiation Syndrome

To illustrate the complexity of treating cases of radiation exposure, let us first outline the symptoms of acute radiation sickness.
Nausea and vomiting would occur in 40 to 50 percent of those
exposed to acute doses in excess of 400 rads, as was observed at
Hiroshima within the 2-kilometer ($1\frac{1}{4}$-mile) radius from the
hypocenter.

More than half of all individuals subjected to these doses would
develop diarrhea, malaise, and lesions of the mucous membranes.
The overall clinical picture would be further complicated by traumatic and thermal injuries, which would lead to an increased incidence of circulatory collapse and shock, even for doses less than
1000 rads. Injuries and burns would increase blood and plasma
loss, gravely complicating the condition of those exposed to
radiation.

The combination of all these factors would create a situation in
which immediate medical care would be needed by enormous
numbers of survivors minutes after the explosion. If they could be
treated at that time, then decisive improvement in their conditions for a certain small percentage could be achieved.

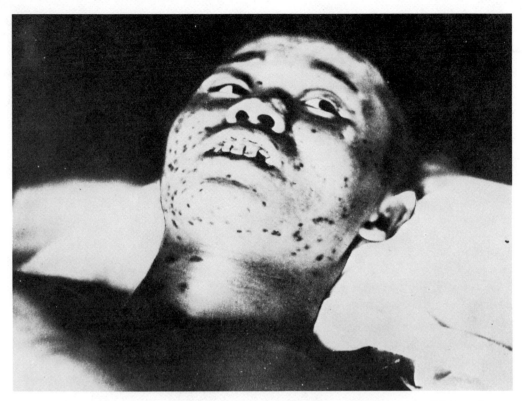

Radiation victim. This 21-year-old man was in a wooden building 1 kilometer (.6 mile) from the hypocenter in Hiroshima on August 6, 1945. His back, right elbow, and abdomen were slashed. On August 18, while receiving treatment, his hair started falling out. On August 29, his gums bled and purple spots of hypodermal bleeding appeared. He was hospitalized on August 30 and developed fever the following day. On September 1, his left tonsil swelled, and the pain in his throat made it difficult to swallow. Bleeding did not stop, and purpuric spots spread over his face and the upper part of his body. On September 2, he lost consciousness and was delirious. He died the following day. This photo was taken two hours before he died. (Kenichi Kimura, U.S. Army returned materials.)

More than one-third of those exposed acutely to 200–600 rads would develop infectious complications due to a lowering of white blood cell levels—pneumonia, enterocolitis, and sepsis. Another one-third would hemorrhage into the skin and gastrointestinal tract. Some would sustain brain or heart hemorrhages, many of which would be fatal. Overall, during the period between two and eight weeks after a nuclear explosion, the cases of radiation exposure alone would demand complex therapeutic procedures

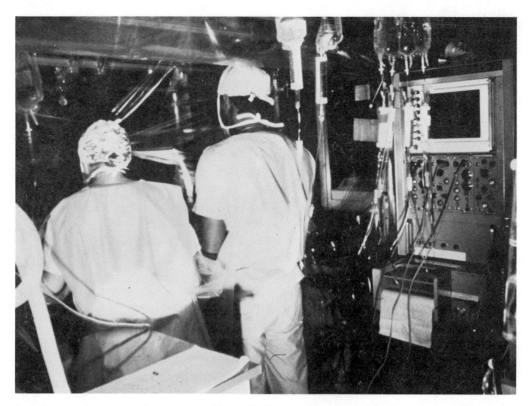

Isolation tent for a severe burn patient. Identical tents would be needed to treat survivors with acute radiation syndrome. (Shriners Burns Institute, Boston.)

available only in a fully equipped modern hospital, carried out on a mass scale and under favorable circumstances. These procedures include the use of expensive antibiotics, transfusions of salt solutions, and massive amounts of blood components (platelets, white cells, and packed red blood cells), requiring an enormous number of donors. In addition, bone marrow transplantation would be crucial in some of those irradiated at high doses. In a nuclear war, few, if any, of these life-saving medical treatments would be available.

Let us look at the likely demand for hospital beds following a nuclear attack. The duration of hospital treatment required for a significant percentage of the injured would be three to four months or more from the time of the explosion. Each of the injured, especially those with multiple injuries, would need continuous medical care, food, and the prolonged application of complex diagnostic methods. Patients with acute radiation syndromes, even without superimposed injuries, are among the most difficult to diagnose and treat. The many complications of their illnesses would require extensive radiologic examinations, laboratory tests, and experienced personnel. In spite of these measures, under the best of conditions, a substantial proportion would die.

Other effects of radiation exposure (see Chapters 7 and 8) include a marked increase in miscarriages, an increase in the frequency of birth defects, a disruption of spermatogenesis and female fertility, epilation, the production of cataracts, damage to the developing brains of children (particularly where there has been both neutron and gamma radiation), and a depression of bone marrow formation of red and white blood cells and platelets.

Because of bone marrow failure, an irradiated individual loses some or all of his ability to fight viral and bacterial infections. This breakdown in immune defenses, along with the disruption of public sanitation, would cause mass outbreaks of communicable disease (dysentery, tuberculosis, hepatitis, and influenza) following a nuclear war.

The consequences of radiation exposure in more resistant organs and tissue take longer periods of time to become apparent. Signs of damage are more difficult to detect but are nevertheless frequent in the range of exposure doses compatible with survival. They include endocrine and metabolic abnormalities and autonomic dysfunctions (dysfunctions of the involuntary nervous system). Those effects are even more likely if radiation exposure is combined with infections due to burns and wounds.

In assessing the consequences of a nuclear strike, we cannot ignore the experience of World War II, which demonstrated that

---

enterocolitis inflammation of the mucous membrane of the small and large intestines.

rad unit of radiation; 450 rads in a short period of time is a lethal dose for approximately 50 percent of the healthy adults exposed.

sepsis presence of various pus-forming and other pathogenic organisms, or their toxins, in the blood or tissues.

239

Homeless and demoralized woman, World War II, 1942. The demoralization of the population surviving a nuclear war would add to the impossibility of delivering adequate medical care. (Tass from Sovfoto.)

World War II: the house of Pavlov and the hospital in Stalingrad, immediately after the battle in February 1943. At least 80 percent of urban hospitals would be destroyed in a nuclear war. (Novosti from Sovfoto.)

when the numbers of injured are large, there is a substantial increase in fatal outcome. In this same context the overwhelming psychological stress would substantially complicate the clinical course and outcome of injuries after a nuclear strike.

Demoralization of the population, with deep depression and inadequate response behavior, would hinder the delivery of timely, adequate medical care. Providing medical assistance would be further complicated by the deaths of 50 to 90 percent of medical personnel; the destruction of 25 to 65 percent of housing; the difficulties in finding and transporting the injured and sick because of roadblocks, raging fires, and the destruction of familiar landmarks; the persistent high levels of radiation; and the disruption of food and water supplies. Just the burial or cremation of the enormous number of corpses would consume a significant portion of the survivors' energy.

241

Tokyo after the March 10, 1945, fire-raid. Piles of corpses, charred beyond recognition, lie in a street. It took the survivors 25 days to collect and dispose of the bodies. In a nuclear attack, the hundreds of thousands of dead and the persistent high levels of radiation could make disposal of the bodies impossible. (Koyo Ishikawa, Tokyo.)

Subclinical Radiation Syndrome and Long-Term Effects

Another question that has been analyzed is that of the long-term consequences of subclinical radiation sickness. Broad extrapolations from clinical and experimental data, using both cases of accidental radiation exposure and the data from Hiroshima and Nagasaki, lead to the following conclusions. Individuals acutely exposed to 100–400 rads experience a depletion of peripheral blood cells of varying duration and degree of expression. The syndrome is characterized by substantial death of the bone marrow with disruption of the division of blood-forming cells and the subsequent mobilization of atypical cells. The radiation creates an increased number of errors in DNA synthesis, reflected by

242

the number of chromosomal aberrations in somatic cells, leading, it is believed, to the development of leukemia.

At Hiroshima and Nagasaki, the increase in leukemia was observed at the highest levels during the first and second five-year periods after the explosions, reaching three to four times the expected incidence. If we extrapolate these cases to the millions who would be exposed to radiation in a nuclear war the number of leukemia cases would be in the tens of thousands. [Editors' note: Chapter 16 calculates that, after a 6500-megaton attack, approximately 45 million survivors in the United States would have been exposed to more than 100 rads acutely. Leukemia cases could then reach 30,000 if the Hiroshima rate of 64 cases per 100,000 survivors applied.] The incidence of other malignant neoplasms, notably of the thyroid, breast, and lung would correspondingly increase.

Among women who were less than 1.5 kilometers from the hypocenter at Hiroshima and Nagasaki, there was a 2.5-fold increase in the incidence of cancer of the thyroid, with a direct relationship to the dose of radiation. Exposure to radioactive iodine that would be present in the post-explosion fallout would increase radiation levels in the thyroids of children and thus increase the risk of tumor development and thyroid dysfunction, compromising normal growth and development. The effects might not become apparent until 14 to 26 years after exposure.

At Hiroshima and Nagasaki higher than normal mortality was observed in children under one year of age born to mothers exposed during the third trimester of pregnancy. This higher mortality was thought to be due, not just to the radiation, but to the mothers being ill during pregnancy and in the early neonatal period. Past experience supports this conclusion. For example, during the siege of Leningrad during World War II, malnutrition and psychological stress were shown to increase the morbidity and mortality of the 80,000 babies born. In 1941–1942, 41 to 60 percent of all pregnancies resulted in premature births, with a 2.4 percent mortality associated with labor and delivery, ten times the pre-war incidence. Babies were shorter, had lower birth weights, and smaller head and chest circumferences. At Hiroshima and Nagasaki, among children born up to ten years after the bomb, there were still cases of delayed skeletal maturation and irregular menstrual cycles as adults. All these effects on children would be greatly magnified following a nuclear war.

Conclusion

A brief analysis of the effects of radiation exposures following a nuclear war has been given in this chapter. We can see from these observations that a nuclear war would lead to death, injury, and disease on a scale that would exceed all previous wars, natural catastrophes, and epidemics. There would be massive numbers of deaths from radiation and the combination of injuries, intolerable changes in the living conditions for all those who survived, and incredible, if not impossible, problems in providing the necessary medical care and the most elementary human needs.

The fate of those at Hiroshima was the result of a bomb that had a total power equivalent to 12.5 kilotons of TNT. We would like to remind the reader that this is an infinitesimally small fraction of today's nuclear arsenal, which totals more than one million times this amount.

Physicians and other scientists from all over the world should join together in the common struggle for peace and the prevention of nuclear war. One of the most important forms of participation in this struggle is the preparation of objective scientific information warning humanity of the possible consequences of a nuclear war. It must be made available to the general public of all countries.

References

Georgievskii, A. S., and Gavrilov. 1975. *Social and Hygienic Consequences of War*. Moscow: Meditsina.

Gladkikh, P. F. 1980. *Health Care in the Besieged Leningrad*. Moscow: Meditsina.

Guskova, A. K. 1977. "Main Results and Prospectives of the Therapeutic Approaches to Acute Radiation Sickness in Humans." In *Medical and Biological Aspects of Radiation Safety*, A. I. Burnazian, ed. Moscow, p. 93.

Guskova, A. K., and Barabanova. 1979. "Radiation Sickness." In *Internist's Reference Manual*, F. I. Komarov, ed. Moscow: Meditsina, p. 487.

Guskova, A. K., and G. D. Baysogolov. 1971. "Radiation Sickness in Man." Moscow: Meditsina.

Guskova, A. K., ed. 1978. *Instructions for Diagnosis, Medical Classification and Therapy of Acute Radiation-Induced Lesions*. Minzdrav.

Handling of Radiation Accidents. 1977. Proceedings of a Symposium, Vienna. 23.2–4.3.

Ilyin, L. A. 1977. *Basics of Body Defense Against Effects of Radioactive Substances.* Moscow: Atomizdat.

Storb, R., and G. W. Santos. 1979. "Bone Marrow Transplantation." *Transplantation Proceedings, 11* (March): 1163–1166.

United Nations Scientific Committee on the Effects of Atomic Radiation. 1977. *Sources and Effects of Ionizing Radiation.* Report to the General Assembly, with annexes. New York: United Nations.

Vishnevski, A. A., and Shriber. 1975. "Multiple Injuries." In *Military Surgery,* p. 60.

Vorobjev, A. I., et al. 1973. "Two Cases of Severe Acute Radiation Sickness." *Terap Arkhiv, 45:* 85.

SECTION V

Nuclear War, 1980's: Environmental and Psychological Consequences

It is essential that not only governments but also the peoples of the world recognize and understand the dangers in the present situation. . . . Removing the threat of a world war — a nuclear war — is the most acute and urgent task of the present day. Mankind is confronted with a choice: we must halt the arms race and proceed to disarmament or face annihilation.

From the Final Documents
United Nations General Assembly
First Special Session on Disarmament
June 30, 1978

18

The Consequences
of Radioactive Fallout

Patricia J. Lindop, M.D., F.R.C.P.,
and Joseph Rotblat, Ph.D., D.Sc.

Radioactive fallout is a unique property of nuclear weapons explosions, but we have almost no actual experience in dealing with its complex effects. The bombs used on Hiroshima and Nagasaki in 1945 were detonated high enough above the ground so that no early fallout occurred, except for "rainout" in a few localities. The atmospheric tests of nuclear weapons were generally carried out in uninhabited areas; the main effect on populations, therefore, was from global fallout. There are persistent reports of military personnel and civilians having been exposed in the vicinity of tests, but no quantitative data have been released. The only fully documented human exposure to local radioactive fallout is the "Bravo" test of March 1, 1954, on Bikini Atoll in the Pacific Ocean. Although the number of persons exposed was small, this incident provided valuable material for estimating the extent and, particularly, the duration of the effects of fallout. However, it hardly gives an idea of the magnitude of the problems that would face the medical profession following the detonation of large numbers of nuclear weapons in a densely populated country. [Editors'

Adapted with permission from *The Final Epidemic: Physicians and Scientists on Nuclear War*, edited by Ruth Adams and Susan Cullen (Chicago: Educational Foundation for Nuclear Science, 1981). The photographs on pages 250, 266, and 267 have been added.

Atomic bomb test at Yucca Flats, Nevada, 1952. Radioactive fallout is produced by surface nuclear explosions carrying particles from the explosion site into the atmosphere and making them radioactive by the fission process. These suspended fallout particles are carried by winds, eventually to fall back to the earth in hours to weeks as local fallout or to remain aloft for months to years and become widely dispersed as global fallout. (U.S. Air Force.)

A major nuclear war ... would be a catastrophe inflicted by
mankind upon itself, by instruments it had itself devised. To
the survivors, if any, the world of today, with all its horrors
and atrocities, would appear in recollection like paradise lost.
The emotional attitudes of men and women in the blasted
world are hard to conceive; they would surely range from
agonizing grief to apathetic despair, with a haunting sense of
terrible guilt at the thought that mankind had squandered and
destroyed its inheritance.

From the Final Summary Documents
Second Congress, International Physicians
for the Prevention of Nuclear War
Cambridge, England
April 1982

note: There is also evidence that people in some counties of Utah
were exposed to radioactive fallout from atmospheric nuclear testing during the period 1951 to 1958, resulting in a marked increase
in the incidence of leukemia among the 10- to 14-year-old age
group (see Lyon et al., 1979).]

Radiation Doses from Fallout

Radioactivity in fallout can expose populations in several ways
and in different time sequences:

- external irradiation by the radioactive cloud as it passes
  overhead
- internal irradiation through the inhalation of radioactive particles in the air
- external irradiation, mainly by the gamma rays from the radioactive substances deposited on the ground
- internal irradiation from eating meat or drinking milk from
  animals that had incorporated such substances, from drinking
  contaminated water, or from eating contaminated crops

In the case of local fallout the external irradiation by gamma-
ray exposure from matter deposited on the ground represents the

Table 1. Dose rates due to gamma rays from fallout at various times after a nuclear explosion.

Time (hours)	Relative Dose Rate (rads/hour)
1	100
2	40
4	15
6	10
12	5.0
24	2.4
36	1.6
48	1.1
72	0.62
100	0.36
200	0.17
500	0.050
1000	0.023

most important hazard. It gives rise to total-body exposure, and the dose rate is proportional to the deposited activity. If all the gamma-ray radioactivity resulting from the detonation of a one-kiloton fission bomb is deposited uniformly over an area of 1 square kilometer (0.4 square mile), then the dose rate (at a height of 1 meter (3 feet) above the ground) is 7500 rads per hour, one hour after the explosion. Allowing for unevenness in terrain, which may cause some of the gamma rays to be absorbed into the ground, and considering that only about 60 percent of the radioactive content of the fallout is deposited as local fallout, the gamma-ray radioactivity from the fallout of a one-kiloton fission bomb yields a dose rate of about 3000 rads per hour at one hour after the explosion. For weapons of other yields, the dose rate increases in proportion to the fission content of the bomb. The actual dose rate is modified by two factors: the decay of the radioactivity with time and the spreading of the fallout with distance.

Time Variation of Dose Rate and Accumulated Dose

Fission products undergo radioactive decay, and therefore the dose rate decreases rapidly with time. Table 1 gives the average dose rates due to gamma rays from early or local fallout, at different times after an explosion. In the table the dose rate at one hour after the explosion is assumed to be 100 rads per hour. If the dose rate at any one time is known, then the dose rates at other times can be obtained simply by proportion. The variation of dose rate with time given in Table 1 is valid only after the fallout is complete at a given place and if no additional debris or material is brought into the location by another explosion and if no material is dissipated because of weather.

The total dose received by a person is obtained by multiplying the dose rate by the time of exposure. Since, however, the dose rate rapidly decreases with time, the calculation of the total dose involves an integration over the relevant values of dose rates. The result of such integration is presented in the figure on p. 254, which gives the accumulated dose starting from one minute after the explosion up to the given time, assuming that the dose rate at one hour is 100 rads per hour. For other values of the dose rate, multiply the values in the figure by the actual dose rate at one hour.

The figure also allows the calculation of the total dose received by a person who enters a given fallout locality at a certain time after the explosion and remains in it for a certain period. This dose is given by subtracting the dose at the time of entry from the dose at the time of exit. The figure also shows that the total accumulated dose tends toward a finite limit, namely 930 rads if the dose rate at one hour is 100 rads per hour. This value, 930 rads, gives the infinity dose, that is, the total dose accumulated starting from one minute after the explosion until an infinite time.

The method of dose calculation outlined above applies only to a single detonation. If the fallout in the given locality is caused by several bombs, exploded at different times, the variation of dose rate with time will be quite different from that given in Table 1, and a single measurement of the dose rate in that locality would not be sufficient to calculate the accumulated dose.

The rapid rate of decay of the early fallout is often used to reassure the population that the danger from fallout radioactivity is over after about two weeks. This reassurance is misleading. What matters is not the dose rate but the accumulated dose, and the latter decreases with time less rapidly than the dose rate. In any case, if the initial fallout in the given area is of high intensity, then the exposure level could be dangerously high for a long time and the area may remain uninhabitable for many years. For example, if the dose rate at one hour is 10,000 rads per hour (which would be the case in many areas targeted by nuclear weapons),

253

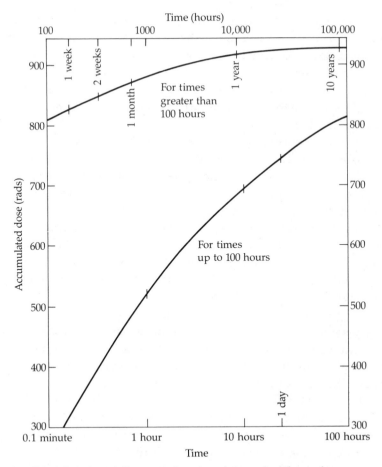

Accumulated dose from fallout as a function of time after the nuclear explosion, assuming that the dose rate is 100 rads per hour at one hour after the explosion. For times up to 100 hours, use the lower curve; for times longer than 100 hours, use the upper curve.

then a person entering the fallout area after one month and then remaining there could accumulate a dose of about 5000 rads in one year. Even if he entered after one year, the dose accumulated during the next year could be about 300 rads, but weathering may reduce this amount.

Distribution of "Early" Fallout

For a given bomb the distance traveled by the fallout particles and the time and location of their deposition are primarily dependent on the speed and direction of the wind. As the particles

254

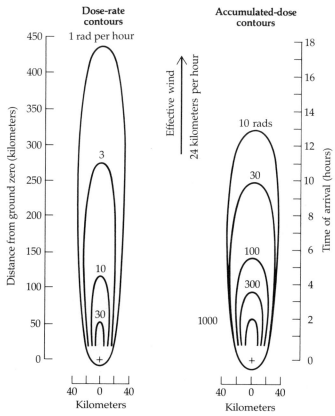

Contours for fallout from a two-megaton bomb (50 percent fission), 18 hours after the explosion.

are carried farther by the wind, they spread over larger areas so that the rate at which the radiation dose may be delivered is rapidly reduced with the distance. This decrease is in addition to the radioactive decay that occurs during the time before the fallout particles reach the ground. Under steady wind conditions, with a constant direction and speed, at any given time after the explosion, radioactivity spreads itself in such a way that the lines joining all points with the same dose rate are cigar-shaped. The figure above shows several dose-rate contours for the fallout from a two-megaton bomb with a 50 percent fission content at 18 hours after the explosion and the contours of the total doses that would result from exposure at these dose rates.

These contours represent a transient situation. At any given place the activity first increases with time, as more fallout reaches

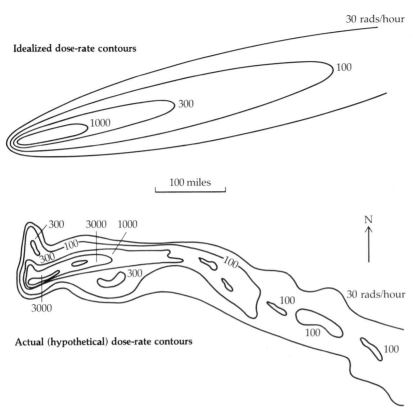

Idealized dose-rate contours

30 rads/hour

100

300

1000

100 miles

300 3000 1000

N

100

300

300

100

3000

30 rads/hour

100

Actual (hypothetical) dose-rate contours

100

100

100

Comparing idealized and hypothetical dose-rate fallout patterns. These dose-rate contours are for a ten-megaton surface burst (50 percent fission) and a wind velocity of 50 kilometers per hour (31 mph). The numbers are dose rates in rads per hour at one hour after the explosion. (From Glasstone and Dolan, 1977.)

a locality and is deposited, but subsequently it decreases, through natural decay. The figure grossly simplifies the situation. In reality the picture is much more complex because many factors may disturb the pattern of the fallout. The wind may change in speed or direction, distorting the shape of the contours. If the airborne particles encounter precipitation, the rain or snow will collect the fallout. Such events may give rise to "hot spots," small areas where a very large amount of radioactivity is deposited. Rain coming down after the fallout may wash away some of the activity. All these phenomena have actually been observed in test explosions.

Table 2.   Dose rates and accumulated doses in an idealized pattern of fallout.

Downwind Distance (km)	Average Time of Deposition (hours)	Reference Dose Rate (rads/hour)		Accumulated Dose (rads)	
		One-Megaton Bomb	Ten-Megaton Bomb	One-Megaton Bomb	Ten-Megaton Bomb
100	4.2	270	1410	780	4100
150	6.2	160	670	420	1770
200	8.3	110	450	270	1110
250	10.4	76	330	180	770
300	12.5	54	260	120	580
350	14.6	42	220	90	470
400	16.7	32	180	66	370
450	18.7	25	160	50	320
500	20.8	20	130	39	250
550	22.9	16	110	30	210
600	25.0	13	98	24	180
650	27.1	12	87	21	160
700	29.2	9.0	74	16	130
750	31.2	7.4	66	13	110
800	33.3	6.3	61	10	100

Thus, the prediction of the hazard due to early fallout is highly unreliable. For example, the figure opposite compares an idealized fallout pattern with a pattern that might actually result from variations in local meteorological and surface conditions. Note that the fallout drifts in a southeast direction rather than to the northeast as anticipated. Also near the explosion there are unexpected low fallout levels, and at other locations downwind, "hot spots" with significantly higher levels—in some areas ten times as high as the idealized contours would predict. Nevertheless, an idealized picture still provides useful information on early fallout. Assuming a wind of constant velocity and direction, and no precipitation, the parameters of the fallout pattern can be computed for bombs of different yields exploded near the surface.

Table 2 contains results of such calculations for one-megaton and ten-megaton thermonuclear bombs with half of the yield due to fission. The average time of deposition of the fallout, the reference dose rates (at one hour after the explosion), and the accumulated infinity doses for persons in the open are presented as a function of downwind distance. (The wind velocity is assumed to

Table 3. Areas covered by given accumulated doses from fallout.

Upper Limit of Accumulated Dose (rads)	Area (square kilometers)	
	One-Megaton Bomb	Ten-Megaton Bomb
1000	900	11,000
800	1,200	14,000
600	1,700	18,000
400	2,600	27,000
200	5,400	52,000
100	10,500	89,000
50	19,000	148,000
24	32,000	234,000
10	53,000	414,000

be 24 kilometers or 15 miles per hour.) If the idealized fallout contours are considered ellipses, then areas can be calculated within which given total doses could be accumulated. Area seems to be a more suitable parameter than distance, since even if meteorological conditions distort the contours, the area covered by fallout is likely to be less affected.

Table 3 shows the result of such area calculations for the one-megaton and ten-megaton explosions considered in Table 2. It gives the areas within which the total accumulated dose could reach the value given in the first column. The areas were calculated to a value of 10 rads, which is approximately the limit of the dose that populations may accumulate in a lifetime from peacetime activities involving exposure to radiation (apart from the natural background and medical procedures). [Editors' note: Ten rads in an area is also the limit for one year's accumulated dose, according to the U.S. Reactor Safety Study (1975), before such areas have to be evacuated.] One ten-megaton bomb could produce a dose near this limit over an area that exceeds the land area of almost every European country.

Factors Affecting Dose

The estimates made in the previous section are theoretical values and represent the maximum gamma-ray doses that a person might receive from local fallout. In practice, the gamma-ray doses are likely to be smaller. Because of the long delay in arrival of the

258

fallout in localities remote from the explosion, people would have time to leave the area threatened with fallout or take shelter. The former is based on the assumptions that uncontaminated areas would be left and that they could be identified. These assumptions may be feasible in the case of one or a few bombs but are unlikely in the aftermath of massive bombing.

Even though properly designed shelters are unlikely to be available for the majority of the population, people may be expected to stay indoors, at least during the early period after the bombing. This again assumes that houses fit for people to live in would be left. Staying indoors would reduce considerably the gamma-ray dose. The dose reduction factor depends among other things on the type of the building, the floor level in a multi-story building, and its proximity to other buildings. On the average, a reduction by a factor of 5 can be assumed. This moderation of the hazard would of course not apply to domestic animals left in the open, or to crops.

On the other hand, certain factors may bring about an increase in the dose and in its biological effect. The dose estimates made previously did not take into account the effects of beta rays, which are emitted by nearly all fission products. Beta rays can contribute to the *external* hazard, if the radioactive materials come into direct contact with the skin, the mucous membranes of the mouth and nose, or the eyes. Owing to the short range of beta rays (a few millimeters in tissue), their action is confined to the superficial layers of the skin, but they may cause beta burns. These start with itching and a burning sensation, and then may develop into weeping, ulcerated lesions, causing much discomfort. These lesions could also temporarily incapacitate potential medical helpers.

The main effect of beta rays is their *internal* hazard, which ensues when a person inhales air containing radioactive particles or ingests such particles with contaminated food or drink. The effects due to inhalation depend markedly on the size of the fallout particles. Large particles descend first and give rise to the highest external fallout doses. But the nose filters out large particles, 5 micrometers or more in diameter; thus, inhalation contributes relatively little to the hazard. The smaller particles that do reach the lung may settle down not only in that organ but also, depending on their chemical form, in other organs, in the bones, thyroid gland, and so on.

Of greater importance is the beta-ray dose delivered via ingestion. While the whole-body dose resulting from the internal deposition of radioactive nuclides is a small fraction of the external dose, the doses to individual organs may be as large as, or even larger than, those from the external gamma rays. For example, in the case of adults in the Rongelap Atoll, the radioactive iodine taken in with contaminated water gave rise to an internal dose to the thyroid gland nearly the same as the external gamma-ray dose; in children the dose to the thyroid was three to eight times larger than the whole-body dose.

The main problem about water, particularly in towns, would be its availability. Supplies are likely to be interrupted by the bombing, through the destruction of equipment, pumping facilities, and pipes. Thus, even if reservoirs are not contaminated, the

---

beta radiation radiation in the form of negatively charged particles, identical to electrons moving at high velocity, which cause cellular damage by contact, either externally on the skin or internally when ingested or inhaled.

gamma radiation electromagnetic radiation originating in atomic nuclei, physically identical to x-rays. Gamma radiation is emitted immediately during the nuclear explosion and over time as fallout. Like x-rays, gamma rays are extremely penetrating; living tissue is essentially transparent to gamma rays.

lymphocyte (or **lymph cell**) white blood cell formed in lymph nodes, involved in cellular immunity and in the production of antibodies.

neutrophil granule-containing, white blood cell that combats infection by ingesting invading organisms.

rad unit of radiation; an acute body-surface dose of 450 rads is lethal to approximately 50 percent of the healthy adults exposed; a bone-marrow dose of 250 rads is lethal to approximately 50 percent of the healthy adults exposed.

triage classification of a large number of casualties into three groups: those who will survive without any medical help, those who will die no matter what treatment they receive, and the priority group of those who will survive only if they receive medical treatment.

---

water will either leak out or will not be available. Lack of piped water will compel people to seek other sources. Rainwater stored in open cisterns in fallout areas will be highly radioactive and unsuitable for drinking until the fallout particles have settled to the bottom. Rain falling on land undergoes a natural process of purification, so that groundwater will be much less contaminated.

Fallout may contaminate food when radioactive rain or dust settles on vegetation. The contamination can be removed in some cases by washing the food with clean water if it is available. Some grain crops, however, trap radioactive particles because of their structures, so decontamination by washing would not be effective. Most food contamination will come through direct assimilation of particles deposited on leaves and shoots of plants and this cannot be easily removed, except by waiting for the short-lived radioactive products to decay. The same applies to milk and meat from animals that have grazed on contaminated grass.

An overall assessment of the effect of internal exposure indicates that for local fallout the hazard is small compared with that from external exposure to gamma rays as far as acute effects are concerned. However, evidence from animal experiments implies that the combination of internal and external exposures may act synergistically, and the lethal dose may be reduced much more than would be expected from the separate effects of internal and external exposures.

Radiation Effects

Any exposure to ionizing radiation may produce a harmful effect, but the type and severity of the effect and the time of its appearance vary considerably. The effects depend primarily on the dose of the radiation. With high doses, the symptoms of exposure are noticed shortly after the exposure. These acute effects have to be distinguished from the long-term effects, which may follow exposures to medium-high and to low doses, and may take different forms, the most prominent being the induction of cancer.

The biological effects of radiation have been the subject of intense study for many years, and it has been claimed that more is known about radiation hazards than about all other environmental or occupational hazards of modern society. Yet, when it comes to the quantitative estimate of the harm to humans from a given dose of radiation, there are very large uncertainties, sometimes amounting to an order of magnitude. These uncertainties apply to both acute and long-term effects.

Doses to Organs in the Body

A person may be exposed either externally (when the source of radiation is outside the body) or internally (when he or she inhales or ingests a radioactive substance). In the case of external exposure, we must distinguish between the amount of radiation reaching the surface of the body and that reaching the various organs inside the body. These two quantities differ because of the attenuation of the intensity of the radiation in passing through the body. The degree of attenuation depends on the properties of the radiation as well as on the size of the body and the depth of the given organ.

When considering the injury caused by exposure to radiation, the dose to the given organ is the determining quantity. However, the radiation doses usually quoted in descriptions of the effects of nuclear explosions (including the previous section of this chapter) refer to the tissue dose measured at the surface of the body (identical to whole-body radiation).

The distinction between surface doses and organ (or midline-tissue) doses is important when the population exposed consists of individuals of different sizes, including children. For example, for the same level of whole-body radiation, an infant will receive a larger dose to the bone marrow than an adult. When combined with the greater intrinsic sensitivity of children to radiation, this can literally mean the difference between life and death. A level of external radiation exposure that gives an adult a reasonable chance of survival is likely to kill a child. Babies and infants will be the first to die after a nuclear attack.

Acute Effects

The symptoms of acute effects appear soon after the exposure to radiation, within one hour or two (or even within minutes after very high doses); but death, if it is the result of the exposure, may not occur for some time. Generally, death from acute effects comes within two months after exposure; but there is evidence from the bombings in Japan that it may come much later, up to several years. By that time deaths from long-term effects (particularly leukemia) begin to occur.

Acute effects may manifest themselves when the whole body, or a large part of it, is exposed in a short time to doses from several tens of rads upward. The early symptoms, such as lack of appetite, nausea, and headaches, are part of the so-called prodromal syndrome. With doses up to 100 rads these symptoms soon disappear and recovery is apparently complete. With increasing dose, mortality increases, reaching 100 percent for a dose of about 500 rads to the bone marrow, although in healthy adults survival is possible even for larger doses if special treatment is provided. Death in the dose range of 100 to 500 rads is mainly due to damage to the blood-forming organs. Larger doses are invariably fatal, death being due to disturbances of the gastrointestinal system and homeostasis. At still higher doses the central nervous system fails.

Radiation Sickness

Soon after exposure to radiation, a person may begin to show acute gastrointestinal symptoms and neurologic effects. This prodromal syndrome is popularly known as radiation sickness. The gastrointestinal symptoms are loss of appetite, nausea, vomiting, diarrhea, intestinal cramps, salivation, dehydration, and loss of weight. The neurologic symptoms include easy fatigability, apathy or listlessness, sweating, fever, headache, and low blood pressure followed by shock. All these symptoms occur only at high doses; with low doses, and during the first 48 hours after the exposure, only some of the symptoms may occur.

Many of these symptoms may also be caused by factors other than exposure to radiation. Nervous tension when war breaks out or crowding in shelters may simulate radiation sickness. Witnessing genuine symptoms of radiation exposure in others may evoke similar symptoms, with epidemic results, particularly in young children (Small and Nicholi, 1982). Without a physical measurement of the radiation dose received, it would be very difficult to distinguish between genuine and spurious reactions.

But even if the dose of radiation is known, it will be impossible in many cases to predict the outcome of the exposure in any individual. Very little is known about the mechanism that underlies the occurrence of the prodromal syndrome, but there is evidence of a large variation among people in their response to an expo-

Table 4.   Radiation doses to midline tissue (in rads) which produce radiation sickness symptoms.

| Symptom | Percentage of Exposed Population | | |
	10%	50%	90%
Anorexia	40 rads	100 rads	240 rads
Nausea	50	170	320
Vomiting	60	215	380
Diarrhea	90	240	390

Source: Langham, 1967, p. 248.

sure. This variation applies to all radiation effects. In a given population there may exist subgroups with a higher than average sensitivity to radiation. The reasons for different sensitivities are unknown; perhaps they include the genetic makeup and the general state of health. In any case, when many people are exposed to the same dose of radiation—in the lower range of the prodromal syndrome—some will show early symptoms of radiation sickness, but others will not. Therefore, to calculate the probability of a given symptom occurring, we have to take a statistical approach and speak of the percentage of an exposed population that will exhibit the symptom.

Table 4 gives the midline-tissue doses for a 10, 50, and 90 percent probability of occurrence of various prodromal symptoms. A dose of 50 rads is likely, for example, to induce nausea in 10 percent of persons exposed; a dose of 215 rads will cause vomiting in half of them; and a dose of 390 rads will cause diarrhea in 90 percent of those exposed as well as all the other symptoms.

Median Lethal Dose (LD$_{50}$)

In the dose range of 100 to 500 rads to the bone marrow, the prodromal syndrome is followed by other clinical symptoms if too many bone-marrow blood-forming cells have been destroyed. These symptoms are hemorrhage under the skin, bleeding in the mouth, and bleeding into internal organs. There is also a greater susceptibility to infection. As in the case of the prodromal syndrome, the response to exposure to a given dose, within the range of 100 to 500 rads, differs from individual to individual. Some will die, usually within six weeks and often with severe emacia-

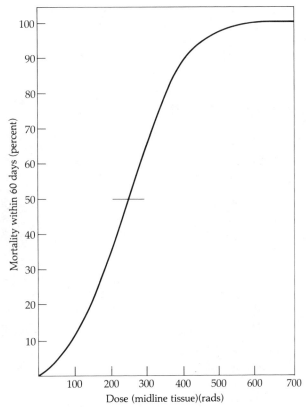

Probability of death from the acute effects of radiation. (From Tobias and Todd, 1974, p. 487.)

tion and delirium. But others will survive, and after some months the number of blood cells will be back to normal, and they will show an apparent recovery (although still at risk from long-term effects). In many cases recovery will take place after a long time and after possible occurrence of such diseases as pancreatitis and liver function changes.

To calculate the probability of death (or survival), the statistical approach is again necessary, but the amount of factual evidence available is very small.

The figure above shows the percentage of persons, out of a large number exposed to a given dose of radiation, who are likely to die as a result of such an exposure. The midpoint of the curve—the dose that gives a 50 percent probability of death within a few

265

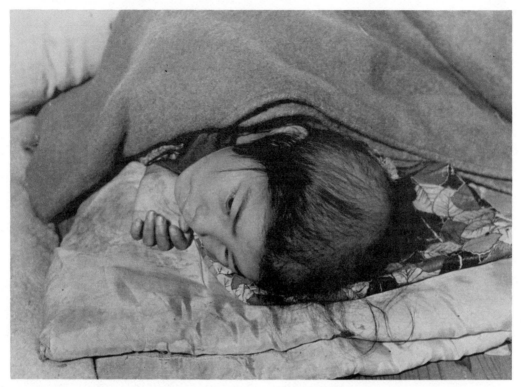

Yo-chan, 12. Trapped under a fallen refrigerator at her home in Hiroshima, about 2 miles from the hypocenter, she lay unconscious for some time after the bombing. Her right thigh was dislocated, and the wounds on her right knee and thigh suppurated. The back of her head was gashed. After about two months her hair began to fall out. She suffered diarrhea and fever—symptoms of radiation disease. October 11, 1945, at a temporary relief station. (Shunkichi Kikuchi, Hiroshima–Nagasaki Publishing Committee.)

weeks after exposure—is called the LD_{50} dose (median lethal dose). LD_{50} values have been measured carefully for many mammals and other living organisms, but for humans this important parameter has been deduced from a small number of observations. The LD_{50} value in the figure, about 250 rads to the bone marrow, may therefore be subject to a large error, and there are suggestions that it is considerably higher.

A remarkable feature of the curve is its steepness. Within a very narrow range of doses, from 100 to 400 rads, the probability of

Yo-chan's mother, Yone-san, 31. Apparently suffering no external injury, she was able to take good care of her daughter. But after about a month, Yone-san began to complain of her bad physical condition. She was hospitalized on October 18, seriously ill, with purple spots of hypodermal bleeding all over her body, her hair falling out, gums bleeding. Frequent coughing made breathing difficult. She soon died. This photo was taken October 11, 1945, at a temporary relief station, Hiroshima. (Shunkichi Kikuchi, Hiroshima–Nagasaki Publishing Committee.)

death increases from 10 to 90 percent. The error in measuring the dose nearly covers this range, so that it would be difficult to say who received a lethal dose. Combining the uncertainties in the LD_{50} value and in the slope of the curve makes any estimate of the number of survivors after an exposure to a bone-marrow dose in the range of 100 to 500 rads very dubious.

The figure applies to adults who received whole-body exposure and who, though not receiving special treatment, would be under care, particularly to avoid infection after the exposure. Conditions of nuclear war would make this case unlikely, and the whole curve might then shift to the left. The extent of the shift is impossible to predict, but any exposure above 100 rads to the marrow might result in death under these circumstances.

Factors Affecting Survival After Acute Exposure

The influence of physical and biological factors on the LD_{50} value is not known sufficiently to be expressed quantitatively with any accuracy, but we can make some general assessments.

The dose rate markedly affects both the occurrence of the prodromal syndrome and the LD_{50} value. From radiotherapy experience we know that protraction of radiation exposure, either by giving it at a low rate or by dividing the dose into a number of fractions separated in time, requires an increase in the total dose to produce the same effect as a dose given in a short time. Radiation damage is thus to some extent repaired in the time between the fractions or during the protracted exposure. In the case of continuous exposure, observations indicate that if the total dose is given in eight days instead of in one day, then all the values in Table 4 for the prodromal symptoms would be doubled.

With regard to the LD_{50} value, it is thought that the curve on page 265 applies if the total dose is delivered within one day or less. Some evidence suggests that the same dose spread over two weeks (the main period of exposure from local fallout) would nearly double the LD_{50} value. At the present time, there is insufficient evidence to establish a clear relationship between the rate of radiation exposure and mortality.

Of the biological factors, the most important is the proportion of the volume of the body exposed to the radiation. Parts of the body (for example, limbs) can be exposed to much higher doses before the acute symptoms occur. In total-body exposure the LD_{50} value also increases if part of the body is shielded from the radiation or receives a smaller dose. The greater the proportion of bone marrow screened from radiation, the greater the chance of survival.

Age at exposure is another relevant factor. Sensitivity to radiation appears to be greater both in the very young and in the old. For these population groups, the LD_{50} value is likely to be smaller than for adults in the middle age group, but the amount of the decrease is unknown.

268

Acute Lung Effects

Death from the acute effects of radiation may also occur as a result of internal exposure, following the inhalation or ingestion of radioactive substances from fallout. Inhalation particularly affects the lungs and may result in death if the dose to the lung tissue is high enough. Long-term effects, such as fibrosis and cancer of the lung, may result from very small doses.

The acute effects ensue both from the direct action of the radiation on the lung walls and from the damaging action on the cells of the lung. The direct radiation affects the permeability of the membrane of the alveoli (air sacs of the lungs), allowing fluids to escape into them. Symptoms are coughing, shortness of breath, and a feeling of "drowning" in lung fluids. The effects on the lung cells include swelling of the alveolar walls, resulting in reduced gas exchange with subsequent hypoxia (deficiency of oxygen); hemorrhage into the alveolar spaces, giving rise to blood-stained sputum; loss of surface-acting secretion, leading to collapse of the alveoli and consolidation of the lung; and loss of the immunological function of the lung, leading to infection and pneumonia.

The causes of death may be heart failure due to hypoxia, pneumonia, or generalized toxemia. The time of death may be some months after the inhalation; it depends on age, environmental conditions, and availability of treatment. The lethal dose to the lungs is 1000 to 2000 rads. A person inhaling a radioactive material is also likely, however, to have internal exposure to other organs as well as external exposure. The combined, possibly synergistic action of the individual exposures makes the prognosis much worse. Under such circumstances smaller doses to the lung may prove lethal.

Estimates of Casualties

Quite large doses of radiation from fallout, in the range of acute lethal effects, could be accumulated at considerable distances from the detonation (see Table 2). Taking into account the variation of the LD_{50} with the dose rate and the time distribution of the dose, and converting from bone-marrow doses to whole-body doses, we can calculate the distance at which a person in the open could accumulate a mean lethal dose. For a one-megaton bomb this distance comes out to be about 120 kilometers (75 miles); for the ten-megaton bomb it is nearly 300 kilometers (186 miles).

From the data in Table 3 the lethal area for acute radiation effects (the area within which the number of survivors would equal the number of fatalities outside the area) can be calculated. For a one-megaton bomb the lethal area is about 1700 square kilometers (650 square miles); for a ten-megaton bomb it is about 18,000 square kilometers (7000 square miles). The latter is over 300 times greater than the lethal area from the initial radiation from the bomb; it is also many times larger than the lethal areas for blast and heat effects.

If we know the population density in the fallout area, we can calculate the number of persons exposed to different doses and the possible casualties. Assuming a population density of 100 persons per square kilometer (the average for Europe), about two million persons might receive a dose from which they would die within a few weeks or months if they were exposed to fallout in the open from a single ten-megaton bomb.

Finally, the effect of war conditions on the chances of recovery from an acute radiation injury must be taken into account. The acute mortality rates calculated in the previous section were based on LD_{50} values applicable in normal conditions. With the lack of food, medical care, and other survival supplies dependent on the social system, which is bound to occur in the wake of a nuclear war, many more people are likely to die following exposure to sublethal doses of radiation. It is difficult to imagine that people will stop eating food contaminated with radioactivity, which they cannot see or smell, when there is nothing else to eat.

Fallout from Bikini

An unintended exposure of a population to the radiation from local fallout occurred in the Marshall Islands, following the Bravo test of March 1, 1954. The explosive yield of this first test of a large thermonuclear device was about 15 megatons. The detonation was near the ground, about 2 meters above a coral reef in Bikini Atoll. Some of the radioactive cloud came down unexpectedly in a long plume in an easterly direction, covering the Marshall Islands, several of which were inhabited by natives and one by U.S. personnel. An area of about 20,000 square kilometers was

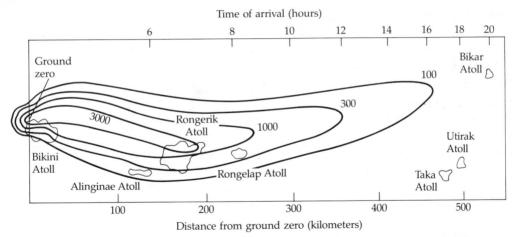

Dose contours 96 hours after the 15-megaton Bravo test explosion. (From Glasstone and Dolan, 1977, p. 436.)

contaminated to such an extent that lethal doses would have been received by persons staying in the open.

The figure above shows the outlines of the atolls and contour lines for several values of total doses that persons might have received during the 96 hours after the explosion. (The accumulated doses to infinity would be nearly double those values.) The bottom and top scales give the distance from ground zero and the time of arrival of the fallout at the given location. Because of an insufficient number of monitoring instruments, the contours were drawn largely by guesswork.

Two days after the test the inhabitants were evacuated. By that time some had received whole-body doses of up to 200 rads (to surface tissue), and the majority of the islanders on Rongelap Atoll had burns from beta radiation. Internal exposure from inhalation and ingestion of radioactive materials (particularly radioactive iodine) with food and water also had occurred. Some inhabitants exhibited symptoms of acute exposure (loss of appetite, nausea, and vomiting). Long-term effects (predominantly thyroid disorders) appeared later.

Almost all the children of the Rongelap Atoll had lesions of the thyroid and had to undergo surgery for the removal of thyroid nodules; later, a number of them showed symptoms of hypothyroidism. Several cases of cancer of the thyroid occurred among the female inhabitants. The population of the Rongelap Atoll islands was allowed to return in 1957, three years after the test; but more than 20 years later, in 1979, the northern islands of the atoll were still declared too radioactive to visit (Johnson, 1980).

The islands of the Bikini Atoll, where testing continued until 1958, remained uninhabited for many years. Vigorous decontamination measures were taken, including the removal of 5 centimeters of the topsoil, before planting new trees. In 1967, the Bikini Atoll was declared habitable, and some islanders returned. However, further geological surveys showed that the radioactivity in the soil was still too high for agriculture, and the atoll was again evacuated. By the end of 1980 some islands of the atoll were declared safe for habitation, but only if 50 percent of the food for the inhabitants was imported (Johnson, 1980).

In addition to the inhabitants of the atolls, a Japanese fishing boat was showered with fallout particles from the Bravo test. The crew was exposed externally from the fallout deposited on the walls and floors of the vessel and on the surfaces of their bodies and internally from the inhaled radioactive materials.

On the first evening their eyes were affected; they suffered from excess tearing and pain in the eyeballs. After two weeks, they developed photophobia (pain on looking at light), edema of the conjunctiva (swelling of the transparent tissue layer covering the eyeball), and acute keratoconjunctivitis (inflammation of the cornea and conjunctiva). Slight opacities of the lens also occurred. Many developed liver damage, from which one died acutely. It is claimed that another died many years later from liver damage. Because their hair and skin were highly radioactive, these men were vigorously scrubbed, and chelating agents were used to remove the contamination. All their hair, including body hair, was shaved off before they were hospitalized. Despite these efforts at decontamination, the men developed acute and long-term skin lesions.

Medical Problems

Physicians must think clearly about what they could do in the event of a nuclear attack. In the post-attack period the first duty would be the care of the dying and the living. The awareness of the enormous numbers of dead or the grotesque destruction of the physical fabric of society, buildings, roads, hospitals, power, and water supplies must not prevent the physician from thinking through what would be required clinically.

Predictions of the radiation dose received by survivors would be necessary for triage. However, because the fallout would not be evenly distributed and radiation monitoring posts are likely to be several miles apart, the dose received by an individual would not be known, even within a considerable error. For estimating the severity of radiation injury, clinical and biological indicators are more important than dose, but laboratory facilities are unlikely to be available for the biological indicators. Therefore, the physician would have to guess the dose from the signs and symptoms.

The presence of nausea and vomiting tends to separate those who have been exposed to a high dose, possibly fatal, from those who have received a low dose, probably nonfatal. Vomiting due to radiation exposure is likely to begin between 20 minutes and 3 hours after exposure, and early onset suggests a high radiation dose. Individual episodes of vomiting may come on suddenly without preceding nausea. Diarrhea is another symptom, with very prompt and explosive diarrhea, particularly if bloody, likely to indicate a fatal dose. The problem is that both vomiting and diarrhea can be caused by emotional stress, which will inevitably be predominant in the horrific aftermath of an attack.

Redness of the conjunctiva may appear fairly promptly after a dose of 150 rads or more, and redness of the skin within a few hours after doses of 500 rads or more, but the doses at which these signs manifest themselves are highly variable. Beta rays cause reddening of the skin at significantly lower doses, which may complicate exposure estimates. This effect, however, will be confined to localized areas of the skin.

The dose to the bone marrow is the primary concern because damage to the blood-forming cells in marrow and lymphatic tissues leads to alterations in blood formation (see the figure on the following page). The changes may vary from mild to lethal. A decrease of lymphocytes occurs promptly, much of it taking place within the first 24 hours after exposure. The amount of this early decrease is one of the best indicators of severity of radiation injury.

In the dose range that damages the blood-cell-forming tissues,

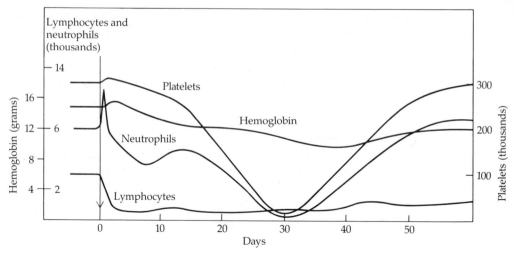

Typical blood values after exposure in the range of 300–400 rads. Damage to blood-forming cells results in changes in the number of other types of cells. (From Andrews, 1980, p. 297.)

there is often an early rise in neutrophils during the first 48 hours, but the numbers fall to fairly low levels at about day 10, followed by a transient, abortive rise around day 15. The absence of an abortive rise is an unfavorable sign. Then there is a steady fall in the count, with a nadir at about day 30. If the patient survives, spontaneous recovery follows beginning in the fifth week.

The time sequence of these changes is not altered much in relation to dose. With the highest dose the onset of pronounced neutrophil and platelet depression occurs earlier, but the time of recovery is only slightly delayed.

The problem is largely to keep the patients alive for about five weeks when marrow recovery will have begun. The two main mechanisms of death after acute radiation exposure are infection and hemorrhage, and in the crowded conditions of shelter life, infection will be a formidable problem. The two tend to develop simultaneously and to be synergistically and rapidly progressive. For example, areas of hemorrhage in the lungs are favorable for bacterial growth, while infectious lesions in the intestinal wall may precipitate bleeding.

In this context it is salutary to recollect the procedure that is considered necessary when dealing with radiation exposure in peacetime. Table 5 sets out a suggested plan for the control of infection following a radiation accident. How could such a plan be even imagined in a post-war situation, when neither the necessary laboratory procedures and personnel nor the isolation facilities would be available? And remember that procedures to prevent infection must begin on day 1 after exposure.

The use of antibiotics and fungicides will have to be carefully considered. Once infection has started in patients receiving doses of 200 rads or more, bone-marrow failure is more rapid and severe. On the other hand, generalized use of fungicides and antibiotics, in populations confined in shelters, is likely to lead to selection of resistant infections and to spread to other members of the group.

A judgment will need to be made between keeping antibiotics for those remaining in shelters—particularly those who might act as sources of infection (for example, the chronic bronchitic) even though they have not received a high radiation dose—or giving antibiotics to those who have already accumulated a dose of 200 rads or who are expected to receive additional doses when going outside. These considerations are much more relevant to planning of post-war survival than are arguments about the value of the LD_{50} or other dose parameters, particularly since it is unlikely that the dose received by a person will be known.

To create the impression that the situation can be handled, civil defense planners use numbers, and even unfounded mathematical formulas, to assess radiation injuries. For example, in the United Kingdom, official civil defense plans consider 150 rads as acceptable exposure if the dose was received in two exposures separated by eight hours (*Nuclear Weapons*, 1980). This number is based on the assumption that there are two stages of response to radiation. The first stage, a relatively quick response to a dose received within about one day, awakens a dominant tolerance to exposure, making the body insensitive to an exposure up to about 150 rads. The second stage, a slow response, consists of the gradual replacement of cells killed by the previous radiation, by virtue of the division and growth of neighboring cells. This stage represents recovery from injury in the normal sense, assuming an estimated recovery rate of at least 10 rads per day.

In accordance with these plans, the wartime emergency dose will be 75 rads as a general rule, except that the persons engaged

Table 5. Suggested plan for treating a potentially lethal radiation syndrome.

Immediately after diagnosis of exposure to 100 rads or more:
- Avoid hospitalizing patient except in sterile environment facility.
- Look for pre-existing infections and obtain cultures of suspicious areas—consider especially carious teeth, gingivae, skin, and vagina.
- Culture a clean-caught urine specimen.
- Culture stool specimen for identification of all organisms: run appropriate sensitivity tests for *Staph. aureus* and Gram-negative rods.
- Treat any infection that is discovered.
- Start oral nystatin to reduce *Candida* organisms.
- Do HLA* typing of patient's family, especially siblings, to select HLA-matched leukocyte and platelet donors for later need.

If granulocyte count falls to less than 1500 per cubic millimeter:
- Start oral antibiotics—vancomycin 500 mg liquid P.O. q. 4 hr.; gentamycin 200 mg liquid P.O. q. 4 hr; nystatin 1 million units liquid P.O. q. 4 hr, 4 million units as tablets P.O. q. 4 hr.
- Isolate patient in laminar flow room or life island.
- Daily antiseptic bath and shampoo with chlorhexidine gluconate.
- Trim fingernails and toenails carefully and scrub area daily.
- For female patients, daily Betadine douche and one nystatin vaginal tablet b.i.d.
- Culture nares, oropharynx, urine, stool, and skin of groins and axillae twice weekly.
- Culture blood if fever over 101°F.

If granulocyte count falls to less than 750 per cubic millimeter:
- In the presence of fever (101°F) or other signs of infection, give antibiotics while awaiting results of new cultures, especially blood cultures. The regimen suggested is ticarcillin 5 g q. 6 hr I.V.; gentamycin 1.25 mg/kg q. 6 hr I.V.
- For severe infection not responding within 24 hours, give supplemental white cells, and if platelet count is low, give platelets from preselected matched donors.
- When cultures are reported, modify antibiotic regime appropriately. Watch for toxicity from antibiotics, and reduce medications as soon as practicable.

When granulocyte count rises to over 1000 per cubic millimeter and is clearly improving:
- Discontinue isolation, antiseptic baths, and antibiotics; continue nystatin for 3 additional days.

---

*HLA: specific protein markers on cell surfaces that identify cells and determine, if transfused, whether they will be accepted or rejected by the host's immunity system.

Source: Andrews, 1980.

in vital tasks may undertake a second period of duty involving an additional 75 rads, provided there is a rest period of eight hours between the two exposures. Persons who have remained in shelters for several days and who have accumulated radiation doses may undertake essential tasks, it is said in these plans, provided that the total exposure does not exceed 150 rads and is acquired over a period not exceeding seven days.

The radiobiological bases for the assumption that the body can repair damage at a rate of 10 rads per day are hard to find and, indeed, are inconsistent with the description of radiation malaise given in *Nuclear Weapons:*

> On average, radiation malaise will result from accumulated doses exceeding 150 r. Its symptoms are fatigue, nausea, indigestion, loss of appetite and, as the dose increases, there may also be vomiting, diarrhoea and the discharge of blood.

Clinical experience shows that although recovery from 150 rads is probable, repeated additional exposure to 10 rads per day is likely to result in severe bone-marrow depression. Indeed, a dose of 10 rads per day over two weeks has been found in radiotherapy to suppress the bone marrow.

Excessive radiation levels will persist for days or even weeks. Thus, in effect, medical care will not be available in many of the shelters in fallout areas. For days or weeks broken limbs will bleed and become infected; a trapped, uninjured person will slowly die from acute radiation sickness—vomiting, diarrhea, and hemorrhage—knowing that no one can come to him. It will not be a question of triage, as applied to conventional disasters. Triage will be accomplished by nature long before doctors and staff can enter the area. When radiation fallout has decreased, however, it will be a question of triage of a different sort: selection for care of those people who are known not to have received a high sublethal radiation dose and who carry no contamination.

On what basis could physicians select for treatment those who had been in areas of heavy fallout? As demonstrated above, it will not be possible in most cases to select those who should receive care because of radiation exposure. Moreover, those who had little external contamination or had been decontaminated by scrubbing and shaving might still have significant internal exposure. In both cases, if the levels were high enough, these exposure victims would have to be turned away, lest they contaminate not just the medical staff and equipment but their own families. Isolation facilities away from family and community would be necessary but not possible. One victim of an inhalation accident in peacetime can be successfully isolated—many cannot.

What would physicians do? Would it be on medical advice that these contaminated people would go back into the fallout areas to carry on rescue work? Would people agree to be measured for radioactive contamination if a high reading—which might be in error by 50 percent or more for gamma radiation or by orders of magnitude for alpha or beta radiation—condemned them to death from radiation?

What criteria would physicians use to decide the radiation dose level at which an individual was not just expendable, but was an insupportable burden on limited resources? These are not decisions that can be abrogated to non-physicians or to the radiation monitoring equipment.

The ubiquitous nature of radioactive fallout, the unpredictability of its distribution, and its persistence render useless any medical planning for dealing with the casualties of a nuclear attack. And underestimating the radiation problems for human beings, let alone for the food and resources on which long-term survival depends, makes civil defense planning for the post-nuclear-attack period an exercise in futility.

References

Andrews, G. A. 1980. "Medical Management of Accidental Total-Body Irradiation." In *The Medical Basis for Radiation Accident Preparedness*, K. F. Hubner and S. A. Fry, eds. Amsterdam: Elsevier/North Holland.

Glasstone, S., and P. J. Dolan. 1977. *The Effects of Nuclear Weapons*, 3rd ed. Washington, D.C.: U.S. Department of Defense and U.S. Department of Energy.

Home Defence Circular (77) 1. 1977. London: D.H.H.S.

Johnson, G. 1980. "Paradise Lost." *Bulletin of the Atomic Scientists*, 36(10):24.

Kumatori, T., T. Ishihara, K. Hirashima, H. Sugiyama, S. Ishii, and K. Miyoshi. 1980. "Follow-up Studies Over a 25-Year Period on the Japanese Fishermen Exposed to Radioactive Fallout in 1954." In *The Medical Basis for Radiation Accident Preparedness*, K. F. Hubner and S. A. Fry, eds. Amsterdam: Elsevier/North Holland.

Langham, W. H., ed. 1967. *Radiobiological Factors in Manned Space Flight.* Washington, D.C.: National Academy of Sciences.

Lyon, Joseph L., M. R. Klauber, J. W. Gardner, and K. S. Udall. 1979. "Childhood Leukemias Associated with Fallout from Nuclear Testing." *New England Journal of Medicine, 300*:397–402.

Nuclear Weapons. 1980. London: Her Majesty's Stationery Office. ISBN 011 340557X.

Small, Gary W., and Armand M. Nicholi. 1982. "Mass Hysteria Among Schoolchildren." *Archives of General Psychiatry, 39* (June):721–724.

Tobias, C. A., and P. Todd. 1974. *Space Radiation Biology and Related Topics.* New York: Academic Press.

U.S. Reactor Safety Study. 1975. WASH–1400 (NUREG–75 014) Washington, D.C.: U.S. Nuclear Regulatory Commission.

19

Changes in Ozone Content from a Nuclear Explosion

Victor N. Petrov, Ph.D.

Ozone, a form of oxygen having three atoms (O_3) instead of two as in the oxygen we breathe (O_2), constitutes a thin layer of the stratosphere, 10 to 20 miles above the earth. Scientists believe that formation of the ozone mantle, hundreds of millions of years ago, permitted the development of higher life forms and the emergence of these organisms onto land. Current theory holds that both major leaps of evolution were made possible by ozone's unique property of absorbing much of the harmful ultraviolet radiation from the sun. Let us consider how nuclear weapons could affect the composition of the ozone layer and what implications these effects could have for human survival following a nuclear war.

Oxides of nitrogen are produced in enormous quantities by atmospheric nuclear explosions. On the average, for each megaton of explosive power, 10^{32} molecules of nitrogen oxides are hurled into the stratosphere. A nuclear war involving the use of 10,000 megatons, approximately 50 percent of the nuclear arsenals expected by 1985 and almost one million times the power of the Hiroshima bomb, would result in the formation of 10^{36} nitrogen oxide molecules, 5 to 50 times more than the stratosphere's total background content.

Nuclear explosion from a height of 12,000 feet. Ten minutes after the explosion, the cloud stem had pushed upward about 25 miles, deep into the stratosphere. The mushroom portion went up to 10 miles and spread for 100 miles. (U.S. Air Force.)

Although the dynamics of the photochemical reactions of the upper atmosphere are extremely complex, and much basic research needs to be done to obtain a more complete understanding of ozone, we know that nitrogen oxides (NO) catalyze the depletion of ozone by the following process:

$$NO + O_3 \rightarrow NO_2 + O_2$$
$$NO_2 + O \rightarrow NO + O_2$$

where the net reaction is

$$O + O_3 \rightarrow 2O_2$$

Atomic oxygen (O) is formed by the action of solar radiation on molecular oxygen (O_2).

Calculations of ozone content are difficult not only because of the complexity of the chemistry but also because of variations in total amounts at different latitudes in each hemisphere and nor-

> **Treat the earth well.**
> **It was not given to you by your parents.**
> **It was loaned to you by your children.**
>
> Kenyan proverb

mal fluctuations at the same latitude over time. Average monthly fluctuations in the northern hemisphere can reach 30 percent. These variabilities may explain some reports that have failed to find a correlation between global ozone content and nuclear explosions. For example, data obtained from the Nimbus-4 Sputnik did not demonstrate a reduction in ozone content following French nuclear explosions of one and two megatons. Other reports, however, including one that relied on the measurement of strontium-90 after nuclear testing and another that followed more precise modeling to determine ozone dynamics, calculated ozone reductions in the northern hemisphere in the range of 1 to 8 percent following atmospheric nuclear explosions.

[Editors' note: The ozone content also can be affected by the introduction of other nitrogen oxides, brought about by the use of agricultural fertilizers, by engine exhaust emissions of supersonic aircraft flying in the stratosphere, and by the accumulation of fluorocarbons in the stratosphere from aerosol cans.]

In spite of all these complexities, Soviet and American scientists are in agreement that a large-scale nuclear war with a 10,000-megaton exchange between the United States and the Soviet Union could destroy from 30 to 70 percent of the ozone layer in the northern hemisphere, and from 20 to 40 percent in the southern hemisphere, with mean reductions of 50 percent and 30 percent respectively (see the figure opposite).

The speed of atmospheric transfer of radioactive products from prior nuclear explosions provides a basis for evaluating the time necessary for ozone restoration. Calculations demonstrate that it would take several years for this restoration to occur by natural atmospheric processes, as the figure shows.

These reductions in ozone content would lead to an increase in harmful ultraviolet radiation (designated UV-B) on the earth's surface by a factor of five or more. What consequences these ultraviolet increases would have for humans and for other living

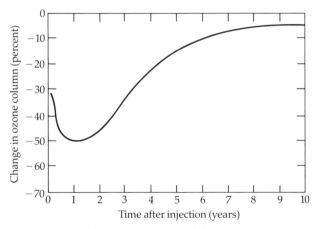

Mean change in ozone column as a function of time after injection of nitrogen oxide from a 10,000-megaton nuclear exchange for the northern hemisphere. (Adapted from National Academy of Sciences, 1975.)

things are not yet clear, the effects on the biosphere being even more complex than the dynamics of ozone chemistry. But it is certain that the effects could be catastrophic, would last for years, and could determine the fate of survivors in the post-nuclear-war world.

Editors' Addendum

We will briefly review possible biological effects brought about by such large increases in ultraviolet radiation exposure. The United States National Academy of Sciences outlined them in its 1975 report, *Long-Term Worldwide Effects of Multiple Nuclear Weapons Detonations.* The following excerpts are from the report.

Although reliable data are meager with regard to effects of UV-B radiation on biological organisms, and the responses are variable and often subtle, the biological implications of elevated UV are far-reaching.

If the upper limits of ozone depletion should be realized (70% reductions), irreversible injury to sensitive aquatic species might occur during the years of increase in UV-B following the detonations.

Many aquatic species appear to have little reserve tolerance to existing solar UV-B. . . . Avoidance by downward migration to a region of lower UV-B intensity may not be possible for some critical components of aquatic ecosystems.

Aquatic ecosystems where significant UV-B penetrates to the full depth of the water (coral reefs, marshes, shallow lakes, clear rivers, etc.) offer only limited possibilities for avoidance of UV-B by sinking.

Photosynthetic organisms . . . cannot move deeper to regions with less visible light to avoid UV-B without some sacrifice in photosynthetic capacity.

That UV-B will produce significant damage to terrestrial ecosystems in the event of multiple nuclear weapons detonations of 10,000-megaton yield is almost certain.

UV-B exposure inhibits plant growth and development . . . reduces photosynthesis, and influences the pollination behavior of insect species.

Agricultural crops (peas, beans, tomatoes, sugar beets, lettuce, and onions) are among the most sensitive plant species to UV radiation. . . . [These] would be severely "scalded" and even killed by a 5–10✕ increase in UV-B.

An increase in UV-B would probably lead to an increased incidence of malignant skin tumors in white-coated or piebald animals. The incidence of "cancer-eye" in Hereford cattle is known to increase with both length and intensity of exposure.

An increase in the skin cancer incidence (in humans) of about 10% (squamous cell carcinoma, basal cell carcinoma, and malignant melanoma), with a range of uncertainty of 3–30% would be experienced at mid-latitudes for about 40 years after the detonations.

The short-term effects of UV-B increase would be severe sunburn in temperate zones and snowblindness in northern regions. For a 50% decrease in ozone . . . a severe sunburn involving blistering of the skin could occur in 10 minutes. Toxic effects due to an increased rate of vitamin D synthesis might occur.

Snowblindness is burning of the cornea and conjunctivae resulting from the near-total reflection of ultraviolet radiation by snow. The syndrome of vitamin D toxicity includes symptoms of weakness, fatigue, loss of appetite, nausea, vomiting, and increased fluid consumption and urination, all presumably secondary to disordered calcium and phosphorus metabolism. Deposits of calcium phosphate crystals throughout the body may cause organ damage, particularly in the kidneys.

Although these effects on biological systems of ozone depletion and resultant UV-B increase are hypotheses not yet fully demonstrated by experimental or empirical data and must presently be considered only speculative, their occurrence would be catastrophic for human survival. To examine the implications of nuclear explosions on the ozone layer, the following questions should be asked:

- What would be the effect on aquatic ecosystems if certain organisms are killed by the increased UV-B? In particular, what would happen to life at all levels of the fresh and salt water food chains if organisms at the base of the chain died?
- What would be the global consequences if major concentrations of phytoplankton, which provide a significant portion of atmospheric oxygen, were killed or had decreased photosynthetic capacity?
- What would be the outcome if terrestrial plants had reduced photosynthesis, or if pollinating insects lost their navigational ability or were blinded by changes in UV-B?
- What would be the effects on post-attack food crops and on livestock because of increased UV-B?
- How would the survivors of a nuclear war be able to remain outside long enough to rebuild their societies or to replant their crops before foodstuffs were depleted, without sustaining disabling skin burns?

These points raise profoundly disturbing questions about the post-nuclear-attack world. Any discussions of surviving a nuclear war must take into account the expected enormous increases in ultraviolet radiation that would result from a depletion of the ozone layer.

References

Alexandrov, E. L., and Y. S. Syedunov. 1979. *Man and Stratospheric Ozone.* Leningrad: Hydrometeoruzdat.

Bauer, E., and F. Gilmore. 1975. "Effect of Atmospheric Nuclear Explosions on Total Ozone." *Reviews of Geophysics and Space Physics, 13* (4).

Chang, I., et al. 1979. "The Atmospheric Nuclear Tests of the 1950's and 1960's: A Possible Test of Ozone Depletion Theories." *Journal of Geophysical Research, 84*(4).

Johnston, H. S. 1977. "Expected Short-Term Local Effect of Nuclear Bombs on Stratospheric Ozone." *Journal of Geophysical Research, 82* (20).

Kondratye, K. Y. 1980. *Radiation Factors of Contemporary Changes in Global Climate.* Leningrad: Hydrometeoruzdat.

National Academy of Sciences. 1975. *Long-Term Worldwide Effects of Multiple Nuclear Weapons Detonations.* Washington, D.C.

20

Survivors of Nuclear War: Psychological and Communal Breakdown

Robert Jay Lifton, M.D.,
and Kai Erikson, Ph.D.

"Scenarios" about fighting, recovering from, or even "winning" a nuclear war tend to be remarkably vague about the psychological condition of survivors. Some simply assume that survivors would remain stoic and begin to rebuild from the ruins in a calm, disciplined way. Others seem to attribute that rebuilding to a mysterious, unseen hand. Usually absent is a reasoned estimate, on the basis of what experience we have, of how people might actually behave. Recently, physicians and other scientists have been making careful projections of the effects of nuclear war, and all raise severe doubts about general claims of recovery.

A 20-megaton bomb, for instance, if detonated over New York City, London, or Leningrad, would vaporize, crush, incinerate, or burn to death almost every person within a radius of 5 or 6 miles from the center of the blast—more than two million people. Within a radius of 20 miles, a million more or so either would die instantly or would suffer wounds from which they could not recover. If the bomb exploded on the ground, countless others who lived miles away, far beyond the reach of the initial blast and searing heat wave, would be sentenced to a lingering death.

But that picture, harsh as it seems, is inadequate even for a limited nuclear war and certainly for a full-scale one. New York

Nagasaki, the day after the bombing. People wander dazed through the ruins 700 meters from the hypocenter. (Yosuke Yamahata, Hiroshima–Nagasaki Publishing Committee.)

City, for example, would be hit by many warheads, as would other cities, industrial centers, and military targets—hundreds of warheads, maybe thousands. Try to imagine 100 million or more people dead, and lethal amounts of radioactivity scattered over huge areas.

And the survivors? Would they panic? Would they help one another? What would they feel and do?

In Hiroshima, survivors not only expected that they too would soon die, they had a sense that *everyone* was dying, that "the world is ending." Rather than panic, the scene was one of slow motion—of people moving gradually away from the center of the destruction, but dully and almost without purpose. They were, as one among them put it, "so broken and confused that they moved and behaved like automatons ... a people who walked in the realm of dreams." Some tried to do something to help others, but most felt themselves to be so much part of a dead world that, as another remembered, they were "not really alive."

288

Those who were able walked silently towards the suburbs and distant hills, their spirits broken, their initiative gone. When asked whence they had come, they pointed to the city and said "that way"; and when asked where they were going, pointed away from the city and said "this way." They were so broken and confused that they moved and behaved like automatons. Their reactions had astonished outsiders who reported with amazement the spectacle of long files of people holding steadily to a narrow rough path, where close by was smooth easy road going in the same direction. The outsiders could not grasp the fact that they were witnessing the exodus of a people who walked in the realm of dreams.

M. Hachiya, M.D.
Hiroshima Diary: The Journal of a Japanese Physician
August 6–September 30, 1945

The key to that vague behavior was a closing off of the mind so that no more horror could enter it. People witnessed the most grotesque forms of death and dying all around them but felt nothing. A profound blandness and insensitivity— a "paralysis of the mind"—seemed to take hold in everyone. People saw what was happening and were able to describe it later in sharp detail, but their minds, they said, were numbed.

Hiroshima and Nagasaki, however, can provide us with no more than a hint of what would happen in the event of nuclear war. A single weapon today can have the power of 1000 Hiroshima bombs, and we have to be able to imagine 1000 of those exploding in the space of a few minutes. Moreover, in the case of Hiroshima and Nagasaki—and this is absolutely crucial—there was still a functioning outside world to provide help.

In a nuclear war, the process of psychic numbing may well be so extreme as to become irreversible.

Imagine the familiar landscape turning suddenly into a sea of destruction: everywhere smoldering banks of debris; everywhere the sights and sounds and smells of death. Imagine that the other survivors are wandering about with festering wounds, broken limbs, and bodies so badly burned that their features appear to be melting and their flesh is peeling away in great raw folds. Imagine—on the generous assumption that your children are alive at all—that you have no way of knowing whether the radiation they have been exposed to has already doomed them.

The suddenness and the sheer ferocity of such a scene would not give survivors any chance to mobilize the usual forms of psychological defense. The normal human response to mass death and profound horror is not rage or depression or panic or mourning or even fear: it is a kind of mental anesthetization that interferes with both judgment and compassion for other people.

In even minor disasters, the mind becomes immobilized, if only for a moment. But in the event of a nuclear attack, the immobilization may reach the point where the psyche is no longer connected to its own past and is, for all practical purposes, severed from the social forms from which it drew strength and a sense of humanity. The mind would then be shut down altogether.

The resulting scene might very well resemble what we usually can only imagine as science fiction. The landscape is almost moon-like, spare and quiet, and the survivors who root among the ruins seem to have lost contact with one another, not to mention the ability to form cooperating groups and to offer warmth and solace to people around them.

In every catastrophe for which we have adequate records, survivors emerge from the debris with the feeling that they are (to use the anthropologist Anthony Wallace's words) "naked and alone . . . in a terrifying wilderness of ruins."

In most cases—and this, too, is well recorded in the literature of disaster—that sense of isolation quickly disappears with the realization that the rest of the world is still intact. The disaster, it turns out, is local, confined, bounded. Out there beyond the periphery of the affected zone are other people—relatives, neighbors, countrymen—who bring blankets and warm coffee, medicines and ambulances. The larger human community is gathering its resources to attend to a wound on its flank, and survivors respond to the attention and the caring with the reassuring feeling that there is life beyond the ruins after all. That sense of communion, that perception that the textures of social existence remain more or less whole, is a very important part of the healing that follows.

None of that would happen in nuclear war.

There would be no surrounding human community, no undamaged world out there to count on.

290

Nagasaki, before noon on August 10, 1945. A mother and child have received boiled rice balls from an emergency relief party. Standing beside the streetcar rails, they seem not even to have the strength to eat. (Yosuke Yamahata, Hiroshima–Nagasaki Publishing Committee.)

No one would come in to nurse the wounded or carry them off to hospitals. There would be no hospitals, no morphine, no antibiotics.

There would be no succor outside—no infusion of the vitality, the confidence in the continuity of life, that disaster victims have always needed so desperately.

Rather, survivors would remain in a deadened state, either alone or among others like themselves, largely without hope and vaguely aware that everyone and everything that once mattered to them had been destroyed. Thus, survivors would experience not only the most extreme forms of individual trauma imaginable but an equally severe form of collective trauma stemming from a rupture of the patterns of social existence.

Virtually no survivors would be able to enact that most fundamental of all human rituals, burying their own dead. The bonds that had linked people in connecting groups would be badly torn, in most cases irreparably, and the behavior of the survivors likely to become muted and accompanied by suspiciousness and extremely primitive forms of thought and action.

Under these conditions, such simple tasks as acquiring food and maintaining shelter would remain formidable for weeks and months, even years. And the bands of survivors would be further reduced not only by starvation but also by continuing exposure to radiation and by virulent epidemics.

For those who managed to stay alive, the effects of radiation might interfere with their capacity to reproduce at all or with their capacity to give birth to anything other than grossly deformed infants. But few indeed would have to face that prospect.

The question so often asked, "Would the survivors envy the dead?" may turn out to have a simple answer. No, they would be incapable of such feelings. They would not so much envy as, inwardly and outwardly, resemble the dead.

References

Erikson, K. 1976. "Loss of Community at Buffalo Creek." *American Journal of Psychiatry,* 133:302–305.

Erikson, Kai. 1977. *Everything in Its Path: Destruction of Community in the Buffalo Creek Flood.* New York: Simon and Schuster.

Hachiya, M. 1955. *Hiroshima Diary: The Journal of a Japanese Physician, August 6–September 30, 1945.* Edited by W. Wells. Chapel Hill: University of North Carolina Press.

Hersey, John. 1946. *Hiroshima.* New York: Knopf. (Bantam Books, 1979.)

Hiroshima–Nagasaki Publishing Committee. 1978. *Hiroshima–Nagasaki: A Pictorial Record of the Atomic Destruction.* Tokyo.

Lifton, Robert Jay. 1967. *Death in Life: Survivors of Hiroshima.* New York: Random House.

Lifton, Robert Jay, and E. Olson. 1976. "The Human Meaning of Total Disaster." *Psychiatry, 39:*1–18.

Nagai, T. 1951. *We of Nagasaki: The Story of Survivors in an Atomic Wasteland.* Translated by I. Shirato and H. B. L. Silverman. New York: Duell.

Office of Technology Assessment. 1979. *The Effects of Nuclear War.* Washington, D.C.: U.S. Congress.

Titchener, J. L. 1976. "Family and Character Change at Buffalo Creek." *American Journal of Psychiatry, 133:*295–299.

Wallace, Anthony F. C. 1956. "Tornado in Worcester." Disaster Study No. 3. Committee on Disaster Studies. Washington, D.C.: National Academy of Sciences—National Research Council.

Epilogue

Howard H. Hiatt, M.D.

"Can anything be done, Doctor?" A familiar question from the patient or his family in the face of an incurable disease or seemingly intractable pain. And never is it justifiable to respond in the negative. Never, because even when nothing else can be done, pain can be relieved, and anxiety can be addressed. Whether or not physicians can offer cure, we can always offer care.

Thus, it is not surprising that people take notice when they hear the medical profession state in chorus and unanimously that nothing could be done to alleviate the physical and psychological effects of nuclear weapons. This, and perhaps only this, is the exception to the "Never."

We know that we could do nothing because we have heard military experts and physical scientists describe the magnitude of the physical catastrophe that would result from a thermonuclear detonation, and we have translated that into effects on the population, on doctors, nurses, and other medical personnel, and on hospitals and other medical facilities. Anyone who is not convinced need only listen to the testimony of the few physicians who were survivors of the bombs in Hiroshima and Nagasaki.

In 1981, the American Medical Association passed a resolution concerning the medical consequences of nuclear warfare. It notes that "in targeted areas, millions would perish outright, including medical and health care personnel. Additional millions would suffer severe injury, including massive burns and exposure to toxic levels of radiation without benefit of even minimal medical care." The resolution recommends that the AMA "inform the President and the Congress of the medical consequences of nuclear war so that policy decisions can be made with adequate factual information."

In 1980 and 1981, a remarkable number of physicians have joined together to address this issue. In educating ourselves and others, we have come to feel that the medical perspective is essential for those charged with responsibility for national security.

The numbers projected as casualties of different types of nuclear attack may have various meaning for the military strategist, but they have only one for the physician. There is talk of "only" two million, or twenty million, dead following a "surgical" attack on U.S. military installations. The strategist may subtract twenty million from 230 million and say that there would still be over 200 million people left. The physician knows what is required to care for *one* seriously burned, or crushed, or radiated patient, and multiplies by however many millions would be injured. This is the perspective that the physician must communicate to the strategist. Whatever its strategic implications, the word "limited," as applied to nuclear war, is meaningless in medical terms.

Many who have listened and have heard our message have joined the crusade for the prevention of nuclear war. Others have not yet been reached. This group includes those who still speak of winning—or surviving—a nuclear war. It also includes those who think that anything else is more important or more urgent than this issue. We must press our political leaders now to stop the production of nuclear weapons. We must urge them to find ways to dismantle in balanced fashion those already deployed by *all* nations. By removing the weapons, the leaders of the nuclear powers would reduce the rapidly increasing likelihood that surviving physicians will one day have to say to those in need: "Nothing can be done."

Appendix

It would take a very strong voice, indeed a powerful chorus of voices, from the outside, to say to the decision makers of the two superpowers what should be said to them:

For the love of God, of your children, and of the civilization to which you belong, cease this madness. You have a duty not just to the generation of the present — you have a duty to civilization's past, which you threaten to render nonexistent. You are mortal men. You are capable of error. You have no right to hold in your hands — there is no one wise enough and strong enough to hold in his hands — destructive powers sufficient to put an end to civilized life on a great portion of our planet. No one should wish to hold such powers. Thrust them from you. The risks you might thereby assume are not greater — could not be greater — than those which you are now incurring for us all.

George F. Kennan
Former U.S. Ambassador to the USSR
October 1980

Summary Proceedings of the First Congress of International Physicians for the Prevention of Nuclear War

Airlie, Virginia
March 20–25, 1981

This document, drafted during the last day of the congress, represents the combined efforts and conclusions of seventy-two physicians from the following twelve countries:

Canada
France
Israel
Japan
the Netherlands
Norway
Sierra Leone
the Soviet Union
Sweden
the United Kingdom
the United States
West Germany

**International Physicians for the
Prevention of Nuclear War, Inc.**
225 Longwood Avenue
Boston, Massachusetts 02115
Telephone (617) 738-9404

Dear Colleagues:

The multiplying stockpiles of nuclear weapons of ever increasing destructiveness threaten humankind with an unimaginable catastrophe. The peoples of North America, the Soviet Union and Europe are held hostage by the accelerating arms race. A war without winners endangers not only human survival but the fragile ecology of our planet.

Physicians charged with responsibility for the lives of their patients and the health of the community must begin to explore a new province of preventive medicine, the prevention of nuclear war.

We gathered here because we do not accept the inevitability of nuclear conflict. We met here because we reject the utilization of technology for nuclear weapons rather than for improving the quality of life. We met here because we do not believe that differences between political systems can be resolved by the use of nuclear weapons. We met here because of our abiding faith in the concept that what humanity creates, humanity can control.

Our aim is to alert physicians world-wide of the mortal peril to the public health. Our hope is that physicians will help educate their communities, for only an aroused and informed citizenry can change the course of events.

Bernard Lown, M.D.
President
International Physicians for the
Prevention of Nuclear War

Airlie, Virginia
March, 1981

Preamble

Nuclear war would be the ultimate human and environmental disaster.

The immediate and long-term destruction of human life and health would be on an unprecedented scale, threatening the very survival of civilization.

The threat of its occurrence is at a dangerous level and is steadily increasing.

Even in the absence of nuclear war, invaluable and limited resources are being diverted unproductively to the nuclear arms race, leaving essential human, social, medical, and economic needs unmet.

For these reasons, physicians in all countries must work toward the prevention of nuclear war and for the elimination of all nuclear weapons.

Physicians can play a particularly effective role because they

1. are dedicated to the prevention of illness, care of the sick and protection of human life;
2. have special knowledge of the problems of medical response in nuclear war;
3. can work together with their colleagues without regard to national boundaries;
4. are educators who have the opportunity to inform themselves, their colleagues in the health professions, and the general public.

The following statements were developed by working groups at the First Congress of International Physicians for the Prevention of Nuclear War, meeting at Airlie, Virginia, March 20–25, 1981.

Predictable and Unpredictable Effects of Nuclear War

The consequences of the nuclear attack on Hiroshima and Nagasaki were disastrous. Yet even they do not serve as adequate precedents for the amount of death and destruction that would follow nuclear warfare today. Given any scenario of a massive nuclear strike in present conditions, the fate of the inhabitants of those two cities would be shared by tens to hundreds of millions of people. Even a single one-megaton nuclear bomb explosion (80 times more powerful than that dropped on Hiroshima) over an urban area would cause death and injury to people on a scale unprecedented in the history of mankind and would present any remaining medical services with insoluble problems. In the event of a major nuclear war there would be thousands of such explosions.

We must distinguish between the immediate and the delayed effects of nuclear war. Among the immediate effects are mass deaths in the first hours, days, and weeks after an explosion. These are caused by the simultaneous effects of blast, heat, and large doses of penetrating radiation. The number of such deaths would be magnified catastrophically by the destruction of buildings, by secondary fires, by disruption of all life-support systems including electric power, communication and transportation, and by the destruction and contamination of the water supply and of foodstocks.

It is difficult for us, even as physicians, to describe adequately the human suffering that would ensue. Hundreds of thousands would suffer third-degree burns, multiple crushing injuries and fractures, hemorrhage, secondary infection, and combinations of all of these. When we contemplate disasters, we often assume that abundant medical resources and personnel will be available. But contemporary nuclear war would inevitably destroy hospitals and other medical facilities, kill and disable most medical personnel, and prevent surviving physicians from coming to the aid of the injured because of widespread radiation dangers. The hundreds of thousands of burned and otherwise wounded people would not have any medical care as we now conceive of it: no morphine for pain, no intravenous fluids, no emergency surgery, no antibiotics, no dressings, no skilled nursing, and little or no food or water. The survivors would envy the dead.

It is known from the Japanese experience that in the immediate aftermath of an explosion, and for many months thereafter, the survivors suffer not only from their physicial injuries—radiation sickness, burns, and other trauma—but also from profound psychological shock caused by their exposure to such overwhelming destruction and mass death.

The problem is social as well as individual. The social fabric upon which human existence depends would be irreparably damaged.

302

Those who did not perish during the initial attack would face serious—even lifelong—dangers. Many exposed persons would be at increased risk, throughout the remainder of their lives, of leukemia and a variety of malignant tumors. The risk is emotional as well as physical. Tens of thousands would live with the fear of developing cancer or of transmitting genetic defects, for they would understand that nuclear weapons, unlike conventional weapons, have memories—long, radioactive memories. Children are known to be particularly susceptible to most of these effects. Exposure of fetuses would result in the birth of children with small head size, mental retardation, and impaired growth and development. Many exposed persons would develop radiation cataracts and chromosomal aberrations.

Delayed radioactive fallout from multiple nuclear detonations would render large areas of land uninhabitable for prolonged periods of time, making it impossible to produce the food upon which the survival of whole populations would depend. Aside from the severe effects in the areas most immediately affected by explosion or local fallout, there would occur effects from both ground and air bursts throughout the world. Fallout would increase the incidence of cancers and of genetic defects in nations and populations far from the targeted areas. These and other effects are difficult to quantify, but it is known that they would occur.

The use of nuclear weapons with an aggregate yield greatly exceeding that of all the explosions (including atomic explosions) in human history poses dangers to the entire planet, and to all of mankind. Among these are profound disruptions of the ecological balance—disturbances to all living organisms, crops, and the atmosphere, with consequences of a nature and magnitude we can only guess at. For example, the release into the atmosphere of large quantities of nitrogen, formed during multiple nuclear explosions, could disturb the ozone layer of the atmosphere, which protects the surface of the earth from the penetrating component of ultraviolet radiation; this would probably cause the death of vegetation and animals and injury to people. In the magnitude, duration, and variety of the dangers it poses to biological and social survival, nuclear war has no precedent in the experience of mankind. The survival of civilized life is at stake.

In one likely and specific scenario that we have considered—an all-out nuclear war between the United States and the Soviet Union in the mid-1980's—it is likely that

1. The population would be devastated. Over 200,000,000 men, women, and children would be killed immediately. Over 60,000,000 would be injured. Among the injured,
   · 30,000,000 would experience radiation sickness,
   · 20,000,000 would experience trauma and burns,
   · 10,000,000 would experience trauma, burns, and radiation sickness.

2. Medical resources would be incapable of coping with those injured by blast, thermal energy, and radiation.
   · 80% of physicians would die.
   · 80% of hospital beds would be destroyed.
   · Stores of blood plasma, antibiotics, and drugs would be destroyed or severely compromised.
   · Food and water would be extensively contaminated.
   · Transportation and communication facilities would be destroyed.

3. Civil defense would be unable to alter the death and devastation described above to any appreciable extent.

4. The disaster would have continuing consequences.
   · Food production would be profoundly altered.
   · Fallout would constitute a continuing problem.
   · Survivors with altered immunity, malnutrition, and unsanitary environment, and severe exposure problems would be subject to lethal enteric infections.

5. A striking increase in leukemia and other malignancies would be observed among long-term survivors. It would be most severe in those who were children at the time of exposure.

6. Profound changes would occur in weather caused by particulates and reduction of atmospheric ozone with attendant alterations in man, animal, and plant species.

7. The effect on adjacent countries is incalculable.

The Role of Physicians in the Post-Attack Period

Considering the known thermal, blast, and radiation effects of a one-megaton thermonuclear explosion over an industrial city of about four million persons, we know that from 200,000 to nearly 500,000 immediate deaths would result, with an additional

400,000 to over 600,000 injured, depending on the nature of the attack.

Instantaneous death would occur as a result of temperatures greater than in the sun itself and from immense blast effects. Physical structures would be converted into unrecognizable rubble and social organization would disintegrate. Many injured would die as a consequence of huge fires and intense radioactive fallout. Neither doctors nor the hospitals in which they work would be spared.

In addition to the dead, there would be the injured—some walking with clothes in shreds and skin peeling in sheets from burns, some trapped in buildings and basements. Many of these would die. Many who were rescued would not survive the crush injuries, multiple fractures, or hemorrhages. Others would die in days or in weeks from burns, traumatic wounds, or radiation exposure.

Many of those injured by a nuclear blast would have combinations of burns, extensive lacerations, and sublethal doses of neutron and gamma radiation. Grave psychological trauma affecting both physician and patient would further aggravate the already severe problems of diagnosis and treatment. These many factors complicate the outcome of therapy and would critically affect medical decisions about who should receive care and who could only be allowed to die with such minimal supportive measures as might be available. Burn and radiation injuries, regardless of other complications, would place the greatest strain on medical personnel and facilities. From the British experience in wartime London, it is estimated that the acute treatment of only 34,000 serious burn cases would require 170,000 health professionals and 8000 tons of supplies.

A city struck by a single one-megaton bomb would find its electrical, water, and food supplies totally disrupted. The techniques of modern medical care would be seriously compromised if not entirely halted. Much of the essential supply of blood, antibiotics, and other materials would be destroyed. A target nation, however, might cope partially with the consequences of having one city struck by a single nuclear bomb. The surviving doctors and other health professionals could respond, supported by help from outside the stricken city, but with severe limitations. The response would fall much below acceptable medical standards.

In peacetime the medical care system can cope successfully with a very small number of the kind of casualties which can be expected in huge numbers from the explosion of a single nuclear bomb. Successful treatments of extensive burns, of crushing injuries, of fractures and lacerations, of perforating wounds of abdomen and thorax, and of sublethal to near-lethal doses of radiation all require the full availability of modern medical technology and the finely developed skills of medical and other support personnel. The medical capacity of any nation would be severely strained, if not for a period overwhelmed, by dealing with the victims of even a single nuclear bomb.

Nuclear war, however, is very likely to involve more than the appalling destruction from a single nuclear bomb, or even a few bombs. With more than 50,000 nuclear weapons in existing stockpiles we must face the prospect of the explosion of hundreds and perhaps thousands of bombs, many possessing hundreds of times the explosive power of those that destroyed Hiroshima and Nagasaki. As tens or hundreds of cities are simultaneously attacked, death and casualties escalate geometrically. The fabric of society would disintegrate and the medical care system, deprived of the facilities developed over the years, would revert to the level of earlier centuries. The surviving walking wounded, physician and layman alike, could only provide what mutual comfort the remnants of their individual humanity would permit. The earth would be seared; the skies would be heavy with lethal concentrations of radioactive particles; and no response to medical needs could be expected from medicine.

The Social, Economic, and Psychological Costs of the Nuclear Arms Race as Related to Health Needs

Preface

The health of mankind is inseparably connected with social, economic, and psychological strengths. The greatest risk of the arms race to health is that it increases the likelihood of nuclear war. Even without such a war, precious human, social, medical, and economic resources are presently diverted unproductively to the nuclear arms race, and this diversion adversely affects health.

Social Costs

Any social undertaking of the magnitude of the arms buildup is bound to affect social structure and social values, regardless of the basis on which that society is built. In particular, activities develop which generate further pressure for more arms and thus establish a dangerous cycle. Moreover, as the scale of arms escalation increases relative to the size of the social institutions and to the strengths of social values, the latter become subverted to, and begin to reflect, the same unproductive and impoverishing priorities and values inherent in the buildup of arms.

Economic Costs

Consideration of economic issues ranges beyond the special expertise of physicians. However, we believe that these issues cannot be completely ignored. The diversion of a major portion of the world's economic resources to armaments increases the likelihood of a nuclear war that would result in death and disability for much of the world's population. This is the ultimate health cost of the arms race and would devastate economic and social organization. The arms buildup weakens the application of existing knowledge, technology, and resources to the prevention and treatment of health problems that currently affect large numbers of the world's population. The arms race increasingly burdens much of the world's population who live in less developed countries. These countries can least afford to use their scarce resources for arms and will suffer grave health and social consequences in doing so. Of greatest importance is that the use of economic resources for armaments diminishes development of knowledge, technology, and manpower that could address global ecological and overpopulation problems. The strains these problems place on the world's limited resources will result, if not resolved, in dire health consequences and, in themselves, increase the likelihood of a nuclear war.

Psychological Costs and Effects

As physicians we can speak about human psychological responses with confidence based on our professional knowledge and experience. Nuclear arms have created a new reality for humanity with profound and widespread psychological effects. The consequences of the use of nuclear weapons defy human comprehension because of the enormity of their destructiveness. This danger grows steadily more acute as nuclear weapons production continues. Studies indicate, among other effects, that living in this threatening context is undermining individual confidence in the possibility of a meaningful personal future. Further studies are needed of the psychological impact of the nuclear arms race upon various groups both in societies which possess nuclear weapons as well as in those that do not.

Living with the possibility of imminent annihilation in a massive nuclear exchange creates an unprecedented threat to individual human beings. Not only does one have to deal with the possibility of one's own agony or sudden death, but one must also confront the potential destruction of all that one loves—humanity itself—forever.

We have identified several psychological mechanisms which can have short adaptive value for the individual in protecting himself from such disturbing emotions as terror and guilt. At the same time these defense mechanisms increase the likelihood that nuclear war will actually occur because they impair the realistic perspectives of those who possess nuclear arms. This prevents the development and use of measures that could take control of the arms race.

1. *Avoidance.* The problem is regarded as too big to handle, too overwhelming, too technical. We leave it to others, to the leaders and the experts, to solve. We become numbed and turn away.

2. *Drawing upon old ways of thinking.* In the face of the terror evoked by an adversary, we seek security, as humanity has traditionally done, through developing ever more dangerous weapons in increasing numbers, and from spurious notions of strength dominated by false concepts of winning and losing. Such thought patterns have become outmoded by the realities of nuclear weapons.

3. *Fear and impulsivity.* The climate of terror created by the superpower confrontation engenders a vicious cycle of fear and mistrust. Fear destroys the capacity for rational thinking and adaptive discrimination and promotes panic-driven, impulsive

308

actions. Such actions provoke fear and similar panic responses in adversaries that further escalate the danger of conflict.

4. *Perceptual distortion.* As a response to threat, regression to archaic thinking patterns occurs, dividing the world into percepts of total goodness and total evil. An adversary comes to be perceived as an enemy that is completely evil, a process which impedes the discovery of areas of common purpose and reduces the ability to deal realistically with actual threat or danger from this or other sources.

5. *Dehumanization.* In order to further justify our hostility toward the adversary, we deny to its leaders and people any human value or worthy motives. The distorted perception of human beings as inanimate objects tends to remove inhibitions against destroying them. The impersonality of graphs and pins on targets, or charts of megatonnage and throw weights (in fact the whole obscene jargon of the nuclear weapons race), destroys not only the appreciation of the humanity of an adversary, but one's own humanity as well.

Concluding Remarks

War is not an inevitable consequence of human nature. War is a result of interacting social, economic, and political factors; it has been a social institution widely used over time to manage conflicts.

To argue that wars have always existed and that this social phenomenon cannot be eliminated ignores history, which has demonstrated a human capacity to change institutions and practices that are no longer useful or are socially destructive. Slavery, cannibalism, dueling, and human sacrifice are among the practices which the human race has recognized to be improper and has abandoned.

The genocidal nature of nuclear weapons has rendered nuclear war obsolete as a viable means for resolving conflict. Because inter-group tensions and conflicts are innate and thus inevitable, effective means for conducting and resolving conflict are indispensable. Human beings have developed and widely used such methods as avoidance/withdrawal, assertive non-violent behavior, unilateral initiative inviting reciprocation, competitive coexistence, negotiation, arbitration, and cooperation.

Rationality and foresight are unique human characteristics which have enabled individuals and groups to override primitive responses, to anticipate future consequences of behavior and to choose courses of action which offer maximal ultimate benefit.

Wars begin in the mind, but the mind is also capable of preventing war.

What Physicians Can Do to Prevent Nuclear War

Review available information on the medical implications of nuclear weapons, nuclear war, and related subjects.

Provide information by lectures, publications, and other means to the medical and related professions and to the public on the subject of nuclear war.

Bring to the attention of all concerned with public policy the medical implications of nuclear weapons.

Encourage studies of the psychological obstacles created by the unprecedented destructive power of nuclear weapons and the ways in which these obstacles prevent realistic appraisal of the dangers of nuclear weapons.

Develop a resource center for education on the dangers of nuclear weapons and nuclear war.

Initiate discussion to develop an international law banning the use of nuclear weapons similar to the laws which outlaw the use of chemical and biological weapons.

Seek the cooperation of the medical and related professions in all countries for these aims.

Encourage the formation in all countries of groups of physicians and committees within established medical societies to pursue the aims of education and information on the medical effects of nuclear weapons.

Establish an international organization to coordinate the activities of the various national medical groups working for the prevention of nuclear war.

An Appeal to the Heads of All Governments and to the United Nations

Advances in technology in the 20th century have benefitted humankind but have also created deadly instruments of mass destruction. The enormous accumulation of these nuclear weapons has made the world less secure. A nuclear conflict would ravage life on earth.

We speak as physicians in the interests of the people whose health we have vowed to protect. The scientific data concerning the medical consequences of the use of such instruments of mass destruction convince us that effective medical care of casualties would be impossible. We therefore urge that elimination of this threat be given the highest priority. No objective is more vital than to preserve the conditions that make possible the continuation of civilized life on earth.

As physicians, we know that the eradication of smallpox, coordinated by the World Health Organization, required intense international communication, cooperation, and dedication. Nuclear war is a far greater threat to humanity. Continuing discussion among the nuclear powers and other countries will be needed to achieve an early cessation of the race to produce these instruments of mass destruction, to prevent their spread, and ultimately to eliminate them.

Respectfully yours,

Participants in the First Congress
of International Physicians for the
Prevention of Nuclear War

Airlie, Virginia
March 23, 1981

An Appeal to the President of the United States of America, Ronald Reagan, and to the Chairman of the Presidium of the USSR Supreme Soviet, Leonid Brezhnev

We, physicians from twelve nations, guided by our concern for human life and health, are well aware of the great responsibility you carry and of the enormous contribution you can make to the prevention of nuclear war.

As physicians and scientists, we have for the past several days reviewed the data on the nature and magnitude of the effects that the use of nuclear weapons would bring. We have considered independently prepared medical and scientific analyses from many sources. Our unanimous conclusions are

1. Nuclear war would be a catastrophe with medical consequences of enormous magnitude and duration for both involved and uninvolved nations.
2. The holocaust would in its very beginning kill tens to hundreds of millions of people. Most of the immediate survivors, suffering from wounds and burns, affected by nuclear radiation, deprived of effective medical care or even water and food, would face the prospect of a slow and excruciating death.
3. The consequences of nuclear war would continue to affect succeeding generations and their environment for an indefinite period of time.

Science and technology have placed the most deadly weapons of mass destruction in the hands of the two nations you lead. This huge accumulation imperils us all. The interests of the present and all future generations require that nuclear war be avoided.

The medical consequences persuade us that the use of nuclear weapons in any form or on any scale must be prevented. To achieve this, we offer you our sincere support.

As physicians, we remember that the eradication of smallpox required intense international communication, cooperation, and dedication. Nuclear war is a far greater threat to humankind. It will require even more intense collaboration among the nuclear powers to achieve an early cessation of the race to produce these instruments of mass destruction.

Respectfully yours,

Participants in the First Congress
of International Physicians for the
Prevention of Nuclear War

Airlie, Virginia
March 23, 1981

An Appeal to the Physicians of the World

Dear Colleagues:

We address this message to you who share our commitment to the preservation of health. Our professional responsibility has brought us together to consider the consequences of the use of nuclear weapons.

We have participated in full and open discussion of the available data concerning the medical effects of nuclear war and its effects on our planet. Our conclusion was inescapable—a nuclear exchange would have intolerable consequences.

Enormous numbers would perish in the first hours and days of a nuclear war. The wounded survivors, burned and affected by nuclear radiation, would face unbearably difficult conditions, without effective medical aid, water, or food. The consequences of a nuclear war would also be disastrous to succeeding generations. A major nuclear exchange would inevitably bring extensive long-term consequences even to countries not directly involved.

No one should be indifferent to the nuclear threat. It hangs over hundreds of millions of people. As physicians who realize what is at stake, we must practice the ultimate in preventive medicine—avoidance of the greatest hazard the world will ever know. Your help is needed in this great endeavor. We urge you

1. to inform yourselves, your colleagues, and the general public about the medical effects of nuclear war;

2. to discuss the medical consequences of nuclear war at meetings of members of medical societies, special symposia, and conferences;

3. to prepare and publish in the medical press and specialized journals articles about medical consequences of the use of nuclear weapons;

4. to speak about medical consequences of nuclear war to medical students and to your community;

5. to use your influence and knowledge to help strengthen the movement of physicians for the prevention of nuclear war.

Respectfully yours,

Participants in the First Congress
of International Physicians for the
Prevention of Nuclear War

Airlie, Virginia
March 23, 1981

Delegates to the First Congress of International Physicans for the Prevention of Nuclear War

Airlie, Virginia
March 20–25, 1981

Affiliations are given for identification purposes only.

Herbert L. Abrams, M.D.   USA
Phillip H. Cook Professor of Radiology, Harvard Medical School

Chief, Department of Radiology, Brigham & Women's Hospital and Sidney Farber Cancer Institute

Regina Armbruster-Heyer, M.D.
West Germany
Coordinator of the Hamburg Physicians Against Nuclear Energy

Member, West German Physicians for the Prevention of Nuclear War

Stanley M. Aronson, M.D.   USA
Former Dean of Medicine,
Brown University Program in Medicine

Professor, Medical Science,
Brown University

A. Clifford Barger, M.D.   USA
Robert Henry Pfeiffer Professor of Physiology, Harvard Medical School

Donald Bates, M.D.   Canada
Chairman, Department of Humanities and Social Studies in Medicine,
McGill University

Thomas Cotton Professor of History of Medicine, McGill University

Sune Bergstrom, M.D.   Sweden
Professor of Biochemistry,
Karolinska Institute

Director, Research Committee, World Health Organization

Robert W. Berliner, M.D.   USA
Dean, Yale University School of Medicine

Professor of Physiology and Medicine,
Yale University School of Medicine

Viola Bernard, M.D.   USA
Clinical Professor Emeritus of Psychiatry, Columbia University College of Physicians and Surgeons

Former Vice President,
American Psychiatric Association

John W. Boag, D.Sc.   United Kingdom
Emeritus Professor of Physics as Applied to Medicine, Institute of Cancer Research, University of London

Past President, British Institute of Radiology

Past President, International Association Radiation Research

Nikolai P. Bochkov, M.D.   USSR
Chief Learned Secretary, Presidium of the USSR Academy of Medical Sciences

Director, Institute of Medical Genetics of the USSR Academy of Medical Sciences

President, National Scientific Society of Medical Genetics

John Burke, M.D.   USA
Helen Andrus Benedict Professor of Surgery, Harvard Medical School

Chief, Trauma Services,
Massachusetts General Hospital

Former Chief, Shriners Burn Institute

Helen M. Caldicott, M.B.A.S., F.R.A.C.P.
USA
Former Instructor in Pediatrics,
Harvard Medical School

Associate in Medicine in Cystic Fibrosis, Children's Hospital Medical Center

President, Physicians for
Social Responsibility

Thomas C. Chalmers, M.D.   USA
President and Dean, Mount Sinai School of Medicine of the City University of New York

Evgueni I. Chazov, M.D. USSR
Member of the USSR Academy of Sciences

Member of the Presidium, USSR Academy of Medical Sciences

Director General, National Cardiological Research Center, USSR Academy of Medical Sciences

President, National Cardiological Society

Eric Chivian, M.D. USA
Staff Psychiatrist, Massachusetts Institute of Technology

John Constable, M.D. USA
Associate Clinical Professor of Surgery, Harvard Medical School

Paul Duchastel, M.D. Canada
President, Association of French-Speaking Physicians of Canada

Jack Fielding, M.D. United Kingdom
Consultant Hematologist,
St. Mary's Hospital, London

Vice Chairman, Medical Campaign Against Nuclear Weapons

Stuart C. Finch, M.D. USA
Professor of Medicine,
Rutgers Medical School

Chief, Department of Medicine,
Cooper Medical Center

Former Director of Research, Radiation Effects Research Foundation, Hiroshima

Jonathan Fine, M.D. USA
Medical Director, North End Community Health Center

Chairman of the Executive Committee, Physicians for Social Responsibility

Alfred P. Fishman, M.D. USA
William Maul Measey Professor of Medicine

Director, Cardiovascular-Pulmonary Division, Department of Medicine, Hospital of the University of Pennsylvania

Jerome D. Frank, M.D. USA
Professor Emeritus of Psychiatry, Johns Hopkins University School of Medicine

Donald S. Gann, M.D. USA
Professor and Chairman, Section of Surgery, Brown University

Surgeon-in-Chief, Department of Surgery, Rhode Island Hospital

Chairman, Committee on Emergency Medical Services, National Research Council

H. Jack Geiger, M.D. USA
Arthur C. Logan Professor of Community Medicine

Director, Program in Health, Medicine and Society, City College–City University of New York

Alfred Gellhorn, M.D. USA
Visiting Professor of Health Policy and Management, Harvard School of Public Health

Former Dean, University of Pennsylvania Medical School

Former Dean, School of Biomedical Education, City College, New York

David S. Greer, M.D. USA
Dean of Medicine, Brown University

Professor of Community Health, Brown University

Angelina K. Guskova, M.D. USSR
Professor of Medical Sciences

Head of Department, Institute of Biophysics of the USSR Ministry of Health

Andrew Haines, M.D. United Kingdom
Epidemiology and Medical Care Unit, Medical Research Council

Arthur H. Hoyte, M.D. USA
Assistant Chancellor for Community Affairs, Georgetown University Medical Center

Assistant Professor in Community & Family Medicine and Obstetrics & Gynecology, Georgetown University School of Medicine

Michito Ichimaru, M.D. Japan
Professor, Department of Internal
Medicine, Atomic Disease Institute, School
of Medicine, Nagasaki University

Leonid A. Ilyin, M.D. USSR
Member of the USSR Academy of
Medical Sciences

Chairman, National Commission for
Radiological Protection

Director, Institute of Biophysics of the
USSR Ministry of Health

Carl J. Johnson, M.D., M.P.H. USA
Associate Clinical Professor of Social and
Environmental Health, University of
Colorado Medical School

Director of Health, Jefferson County
Health Department

John Karefa-Smart, M.D. Sierra Leone
Chairman-elect, International Health
Section, American Public
Health Association

Supervisor, Medical Programs, Howard
University Medical School

Former Assistant Director General,
World Health Organization

Dieter Koch-Weser, M.D. USA
Associate Dean of International Programs,
Harvard Medical School

Howard Kornfeld, M.D. USA
Board of Directors, Physicians for
Social Responsibility

Einar Kringlen, M.D. Norway
Professor and Director, Institute of
Behavioral Sciences in Medicine,
University of Oslo

Chairman, Norwegian Information
Committee on Defense and
National Security

Mikhail I. Kuzin, M.D. USSR
Member of the USSR Academy of
Medical Sciences

Vice President, National Surgical Society

Former Dean, First Moscow
Medical School

Director, Vishnevsky Institute of Surgery
of the USSR Academy of Medical Sciences

Alexander Leaf, M.D. USA
Chairman and Ridley Watts Professor,
Department of Preventative Medicine
and Clinical Epidemiology,
Harvard Medical School

Former Jackson Professor of
Clinical Medicine

Former Chief of Medical Services,
Massachusetts General Hospital

Etienne LeBel, M.D. Canada
Professor and Chairman, Department of
Nuclear Medicine and Radiobiology,
Sherbrooke Medical School

Robert Jay Lifton, M.D. USA
Foundations' Fund Research Professor of
Psychiatry, Yale University

Patricia Lindop, M.D., D.Sc., F.R.C.P.
United Kingdom
Professor of Radiobiology, Department of
Radiobiology, Medical College of
St. Bartholomew's Hospital

Bernard Lown, M.D. USA
Professor of Cardiology, Harvard School
of Public Health

John E. Mack, M.D. USA
Professor of Psychiatry and Chairman
of the Executive Committee of the
Departments of Psychiatry,
Harvard Medical School

Pulitzer Prize Winner

H. Marcovich, M.D., Ph.D. France
Professor, Pasteur Institute, Director of
Research, National Center for
Scientific Research

Jules H. Masserman, M.D. USA
Honorary Life President,
World Association for Social Psychiatry

Professor Emeritus and Former Chairman
of Psychiatry and Neurology,
Northwestern Medical School, Chicago

Past President, American
Psychiatric Association

316

Roy Menninger, M.D. USA
President, The Menninger Foundation

Henri Mollret, M.D. France
Department of Epidemiology,
Pasteur Institute

Martin C. Moore-Ede, M.D., Ph.D. USA
Assistant Professor of Physiology,
Harvard Medical School

James E. Muller, M.D. USA
Assistant Professor of Medicine,
Harvard Medical School

Associate in Medicine, Brigham and
Women's Hospital

Paul F. Muller, M.D. USA
Assistant Dean, Indiana University
Medical Center

Medical Director, St. Vincent's Hospital,
Indianapolis

Henry Neufeld, M.D. Israel
Professor of Cardiology,
Tel Aviv University

Takeshi Ohkita, M.D. Japan
Professor and Director, Research Institute
of Nuclear Medicine & Biology,
Hiroshima University

David L. Pearle, M.D. USA
Associate Professor of Medicine and
Pharmacology, Division of Cardiology,
Georgetown University School
of Medicine

Victor N. Petrov, Ph.D. USSR
Doctor of Physics and
Mathematical Sciences

Department Head, Institute of
Applied Geophysics

Valentin I. Pokrovski, M.D. USSR
Corresponding Member of the Academy of
Medical Sciences

Director, Institute of Epidemiology of the
USSR Ministry of Health

Vice President, National Scientific Society
of Epidemiologists

Kenneth Rogers, M.D. USA
Professor and Chairman, Department of
Community Medicine, University of
Pittsburgh School of Medicine

Rita R. Rogers, M.D. USA
Clinical Professor of Psychiatry,
UCLA School of Medicine

Chief, Division of Child Psychiatry,
Harbor–UCLA Medical Center

Chairman, Task Force on Psycho-Social
Impact of Nuclear Advances,
American Psychiatric Association

Jonas E. Salk, M.D. USA
Founding Director and Resident Fellow,
Salk Institute for the Biological Sciences

E. Martin Schotz, M.D. USA
International Committee, Physicians for
Social Responsibility

Mikhail G. Shandala, M.D. USSR
Corresponding Member of the USSR
Academy of Sciences

Director General, Kiev Research Institute
of General and Communal Hygiene of the
Ukrainian Ministry of Health

Vice President, Ukrainian Scientific
Society of Hygienists

Evgueni V. Shmidt, M.D. USSR
Member of the USSR Academy of
Medical Sciences

Director, Institute of Neurology of the
USSR Academy of Medical Sciences

Honorary President, National Scientific
Society of Neuropathologists
and Psychiatrists

Vice President, World Federation
of Neurologists

Honorary Member, American
Neurological Association

Naomi Shohno, Ph.D. Japan
Professor, Hiroshima Jogakuin College

Chairman, Hiroshima Society for the
Study of Nuclear Problems

Frank Sommers, M.D., F.R.C.P.   Canada
Department of Psychiatry,
University of Toronto

President, Physicians for Social
Responsibility (Canada)

James Titchener, M.D.   USA
Professor of Psychiatry, University of
Cincinnati College of Medicine

Nikolai N. Trapeznikov, M.D.   USSR
Member of the USSR Academy of
Medical Sciences

Deputy Director General, National
Oncological Scientific Center of the USSR
Academy of Medical Sciences

Vice President, National Cancer Society

Vladimir B. Tulinov   USSR
Senior Researcher, Institute of U.S. and
Canadian Studies of the USSR Academy
of Sciences

Marat E. Vartanyan, M.D.   USSR
Corresponding Member of USSR
Academy of Sciences

Deputy Director, Institute of Psychiatry of
the USSR Academy of Medical Sciences

Vice President, National Society of
Medical Genetics

William Verheggen, M.D.
The Netherlands
President, The Netherlands Medical
Association for the Prevention of War

Everhard Weber, M.D.   West Germany
Coordinator of Hamburg Physicians
Against Nuclear Energy

Claude E. Welch, M.D.   USA
Clinical Professor of Surgery Emeritus,
Harvard Medical School

Senior Surgeon,
Massachusetts General Hospital

Emil Wennen, M.D.   The Netherlands
Secretary, The Netherlands Medical
Association for the Prevention of War

Evgueni A. Zherbin, M.D.   USSR
Professor

Director, Leningrad Central Research
Institute of Roentgenology and Radiology
of the USSR Ministry of Health

Vice Chairman of the Board, National
Scientific Society of Roentgenologists
and Radiologists

**International Physicians for the
Prevention of Nuclear War, Inc.**
225 Longwood Avenue
Boston, Massachusetts 02115
Telephone (617) 738-9404

318

Suggested Readings

Hiroshima and Nagasaki

Barnaby, Frank C. 1977. "Hiroshima and Nagasaki: The Survivors." *New Scientist* (London), 75 (August 25):472–475.

Children of Hiroshima. 1980. Tokyo: The Publishing Committee for *Children of Hiroshima.* Distributed by Taylor & Francis Ltd., 4 John St., London WC1 N2ET.

The Committee for the Compilation of Materials on Damage Caused by the Atomic Bombs in Hiroshima and Nagasaki. 1981. *Hiroshima and Nagasaki: The Physical, Medical and Social Effects of the Atomic Bombings.* Translated by E. Ishikawa and D. L. Swain. New York: Basic Books.

Finch, Stuart C. 1979. "The Study of Atomic Bomb Survivors in Japan." *American Journal of Medicine,* 66:899–901.

Hachiya, M. 1955. *Hiroshima Diary: The Journal of a Japanese Physician, August 6–September 30, 1945.* Edited by W. Wells. Chapel Hill: University of North Carolina Press.

Hersey, John. 1946. *Hiroshima.* New York: Knopf.

Hiroshima–Nagasaki Publishing Committee. 1978. *Hiroshima–Nagasaki: A Pictorial Record of the Atomic Destruction.* Available from the Foundation for International Understanding, 18 Fairview Ave., Arlington, Mass. 02174.

Japanese Broadcasting Corporation (NHK), ed. 1977. *Unforgettable Fire—Pictures Drawn by Atomic Bomb Survivors.* New York: Pantheon Books.

Lifton, Robert Jay. 1967. *Death in Life: Survivors of Hiroshima.* New York: Random House.

Nagai, T. 1951. *We of Nagasaki: The Story of Survivors in an Atomic Wasteland.* Translated by I. Shirato and H. B. L. Silverman. New York: Duell.

Nakazawa, Keiji. 1978. *Barefoot Gen (Hadashi No Gen).* Translated by Project Gen. Tokyo. Available from the War Resisters League, 339 Lafayette St., New York, N.Y. 10012.

Consequences of Nuclear Weapons and Nuclear War

Drell, Sidney D., and Frank von Hippel. 1976. "Limited Nuclear War." *Scientific American,* 235 (November):27–37.

Ervin, F. R., J. B. Glazier, S. Aronow, D. Nathan, R. Coleman, N. C. Avery, S. Shohet, and C. Leeman. 1962. "The Medical Consequences of Thermonuclear War—I. Human and Ecologic Effects in Massachusetts of an Assumed Thermonuclear Attack on the United States." *New England Journal of Medicine,* 266:1127–1137.

Federation of American Scientists (FAS). 1981. *Effects of Nuclear War. Journal,* 34 (February). Available from FAS, 307 Massachusetts Ave. NE, Washington, D.C. 20002.

Feld, Bernard T. 1976. "Consequences of Nuclear War." *Bulletin of the Atomic Scientists,* 32 (June).

Fetter, Steven A., and Kosta Tsipis. 1981. "Catastrophic Releases of Radioactivity." *Scientific American,* 244 (April):41–47. Offprint 3101.

Geiger, H. Jack. 1980. "Addressing Apocalypse Now: The Effects of Nuclear Warfare as a Public Health Concern." *American Journal of Public Health,* 70(9):958–961.

Glasstone, Samuel, and Philip J. Dolan, eds. 1977. *The Effects of Nuclear Weapons,* 3rd ed. Washington, D.C.: U.S. Department of Defense and U.S. Department of Energy.

Katz, Arthur M. 1982. *Life After Nuclear War: The Economic and Social Impacts of Nuclear Attacks on the United States.* Philadelphia: Ballinger.

Lewis, Kevin N. 1979. "The Prompt and Delayed Effects of Nuclear War." *Scientific American,* 241 (July):35–47. Offprint 3051.

Mark, J. Carson. 1976. "Global Consequences of Nuclear Weaponry." *Annual Review of Nuclear Science,* 26:51–87.

Medical Campaign Against Nuclear Weapons (MCANW) and the Medical Association for the Prevention of War (MAPW). 1982. *The Medical Consequences of Nuclear Weapons.* Available from MCANW, 239 Tenison Rd., Cambridge CB1 2DG, England.

National Academy of Sciences. 1975. *Long-Term Worldwide Effect of Multiple Nuclear Weapons Detonations.* Washington, D.C.

Nuclear Weapons: Report of the Secretary General of the United Nations. 1981. Brookline, Mass.: Autumn Press.

Office of Technology Assessment of the U.S. Congress. 1979. *The Effects of Nuclear War.* Washington, D.C.

Report of the Secretary General of the United Nations. 1968. "Effects of the Possible Use of Nuclear Weapons and the Security and Economic Implications for States of the Acquisition and Further Development of These Weapons." New York: United Nations.

Rotblat, Joseph. 1981. *Nuclear Radiation in Warfare.* Stockholm International Peace Research Institute. Cambridge, Mass.: Oelgeschlager, Gunn & Hain.

Royal Swedish Academy of Sciences. 1982. "Nuclear War: The Aftermath." *Ambio, 11*(2, 3). Elmsford, N.Y.: Pergamon Press.

"Short and Long Term Health Effects on the Surviving Population of a Nuclear War." 1980. Hearing before the Subcommittee on Health and Scientific Research of the Committee on Labor and Human Resources, United States Senate, 96th Congress, second session, June 19.

U.S. Arms Control and Disarmament Agency. 1979. *The Effects of Nuclear War.* Washington, D.C. (April).

Physicians and Nuclear War

Adams, Ruth, and Susan Cullen, eds. 1981. *The Final Epidemic—Physicians and Scientists on Nuclear War.* Chicago: Educational Foundation for Nuclear Science.

Adams, Ruth, and Susan Cullen, eds. 1982. *De Fatale Epidemie (The Final Epidemic),* with chapters by William Verheggen. Translated by Anne Boermans. Amsterdam: Meulenhoff Informatiev. Published for Nederlandse Vereniging voor Medische Polemologie (Netherlands Medical Association for the Study of War).

American College of Physicians. 1982. "The Medical Consequences of Radiation Accidents and Nuclear War." April 16. Available from American College of Physicians, 4200 Pine St., Philadelphia, Penn. 19104.

American Medical Association. 1981. "Physician and Public Education on the Medical Consequences of Thermonuclear Warfare." Report DD of the AMA Board of Trustees (December).

Caldicott, Helen. 1979. *Nuclear Madness: What You Can Do.* New York: Bantam Books.

Chazov, Evgueni I., Leonid A. Ilyin, and Angelina K. Guskova. 1982. *The Danger of Nuclear War—Soviet Physicians' Viewpoint.* Moscow: Novosti Press Agency.

Frechette, Alfred L. 1982. *Nuclear Weapons—A Public Health Concern.* Available from the Massachusetts Department of Public Health, 600 Washington St., Boston, Mass. 02111.

Hiatt, Howard. 1980. "Preventing the Last Epidemic." *Journal of the American Medical Association, 244:*2314–2315.

Hiatt, Howard. 1981. "Preventing the Last Epidemic II." *Journal of the American Medical Association, 246:*2035–2036.

Lown, Bernard. 1981. "Physicians and Nuclear War." *Journal of the American Medical Association, 246*(20, November 20).

Lown, Bernard, E. Chivian, James Muller, and Herbert Abrams. 1981. "The Nuclear Arms Race and the Physician." *New England Journal of Medicine, 304:*26–29.

Proceedings of the Medical Association for the Prevention of War (MAPW), 3, Part 5, March 1981. Available from MAPW, 57B Somerton Rd., London NW2 1RU.

Sidel, Victor W., H. Jack Geiger, and B. Lown. 1962. "The Physician's Role in the Post Attack Period." *New England Journal of Medicine, 266*(22): 1137–1145.

Psychological Aspects of Nuclear War

Dumas, Lloyd. 1980. "Human Fallibility and Weapons." *Bulletin of the Atomic Scientists, 36* (November):15–20.

Frank, Jerome, D. 1979. "When Fear Takes Over." *Bulletin of the Atomic Scientists, 35:*24–26.

Frank, Jerome D. 1980. "The Nuclear Arms Race—Sociopsychological Aspects." *American Journal of Public Health, 70:*950–952.

Frank, Jerome D. 1982. *Sanity or Survival in the Nuclear Age,* 2nd ed. With Introduction by James Muller. New York: Random House.

Lifton, Robert Jay. 1979. *The Broken Connection.* New York: Simon and Schuster.

Lifton, Robert Jay. 1980. "The Prevention of Nuclear War." *Bulletin of the Atomic Scientists, 36* (October):38–43.

Mack, John E. 1981. "Psychosocial Effects of the Nuclear Arms Race." *Bulletin of the Atomic Scientists, 37* (April):18–23.

Psychiatric Aspects of the Prevention of Nuclear War. 1964. Formulated by the Committee on Social Issues, Group for the Advancement of Psychiatry, 5, Report 57 (September). Available from GAP, 104 E. 25th St., New York, N.Y. 10010.

Psychosocial Aspects of Nuclear Developments. 1982. Washington, D.C.: American Psychiatric Association. Available from APA Publication Sales, 1700 18th St. NW, Washington, D.C. 20009.

Weapons Systems and Arms Control

Aldridge, R. C. 1978. *The Counterforce Syndrome: A Guide to U.S. Nuclear Weapons and Strategic Doctrine.* Washington, D.C.: Institute for Policy Studies.

The Boston Study Group. 1979. *Winding Down: The Price of Defense.* San Francisco: W. H. Freeman and Company.

Epstein, William. 1975. "The Proliferation of Nuclear Weapons." *Scientific American, 232* (April):18–33.

Kaplan, Fred M. 1978. "Enhanced-Radiation Weapons." *Scientific American, 238* (May):44–51. Offprint 3007.

Kendall, Henry W. 1979. "Second Strike." *Bulletin of the Atomic Scientists, 35* (September):32–37.

Kistiakowsky, George. 1978. "The Folly of the Neutron Bomb." *Bulletin of the Atomic Scientists, 34* (September):25–29.

Mandelbaum, M. 1979. *The Nuclear Question: The United States and Nuclear Weapons, 1946–1976.* Cambridge: Cambridge University Press.

Progress in Arms Control? Readings from Scientific American. 1979. Introductions by Bruce M. Russett and Bruce G. Blair. San Francisco: W. H. Freeman and Company.

Scoville, Herbert. 1981. *MX—Prescription for Disaster.* Cambridge, Mass.: MIT Press.

Stockholm International Peace Research Institute. 1981. *World Armaments and Disarmament: SIPRI Yearbook 1981.* Cambridge, Mass.: Oelgeschlager, Gunn & Hain.

Civil Defense

Aspin, Les. 1979. "Civil Defense—The Mineshaft Gap." *Congressional Record*, January 15, pp. E26–E35.

Chivian, Eric. 1981. "Ten Assumptions of Civil Defense Plans for Nuclear War." 31st Pugwash Conference, Banff, Alberta, Canada, August 28–September 2.

Director of the Central Intelligence Agency. 1978. "Soviet Civil Defense." Washington, D.C.: U.S. Central Intelligence Agency (July).

Final Report of Working Group VII—Civil Defense. 1982. Second Congress of International Physicians for the Prevention of Nuclear War. Cambridge, England (April). Available from IPPNW, 225 Longwood Ave., Boston, Mass. 02115.

Kaplan, Fred. 1978. "The Soviet Civil Defense Myth." *Bulletin of the Atomic Scientists*, Part I (March):14–20; Part II (April):41–48.

Kincaid, William. 1978. "Repeating History: The Civil Defense Debate Renewed." *International Security* (Winter).

Liederman, P. Herbert, and Jack H. Mendelson. 1962. "Some Psychiatric Considerations in Planning for Defense Shelters. *New England Journal of Medicine,* 266:1149–1154.

Piel, Gerard. 1963. "The Illusion of Civil Defense." In *The Fallen Sky,* Saul Aronow et al., eds. New York: Hill and Wang, pp. 57–73.

Tucker, Anthony, and John Gleisner. 1982. *Crucible of Despair: The Effects of Nuclear War.* London: The Menard Press.

U.S. Arms Control and Disarmament Agency. 1978. "An Analysis of Civil Defense in Nuclear War." Washington, D.C. (December).

"United States and Soviet Civil Defense Programs." 1982. Hearings before the Subcommittee on Arms Control, Oceans, International Operations and Environment of the Committee on Foreign Relations, United States Senate, 97th Congress, second session, March 16 and 31.

Weinstein, John M. 1982. "Soviet Civil Defense and the U.S. Deterrent." *Parameters, Journal of the U.S. Army War College,* 12 (March):70–83.

Yegorov, P. T., I. A. Shlyakhov, and N. I. Alabin. 1970. *Civil Defense: A Soviet View.* Translated and published under the auspices of the U.S. Air Force.

General Readings

Calder, Nigel. 1981. *Nuclear Nightmares: An Investigation into Possible Wars.* New York: Penguin Books.

Griffiths, Franklyn, and John C. Polyanyi, eds. 1979. *The Dangers of Nuclear War.* Buffalo, N.Y.: University of Toronto Press.

Ground Zero. 1982. *Nuclear War: What's in It for You?* New York: Pocket Books.

Kistiakowsky, George, George Rathjens, Paul Doty, Richard Garwin, and Thomas Schelling. 1975. "Nuclear War by 1999?" *Harvard Magazine* (November).

Nuclear War in Europe. Report of the First Conference on Nuclear War in Europe, Groningen, The Netherlands, April. Available from the Center for Defense Information, 303 Capital Gallery West, 600 Maryland Ave. SW, Washington, D.C. 20024.

Schell, Jonathan. 1982. *The Fate of the Earth.* New York: Knopf.

Sivard, Ruth L. 1980. "World Military and Social Expenditures." Leesburg, Va.: World Priorities. Available from World Priorities, Inc., Box 1003, Leesburg, Va. 22075.

Thompson, Edward P., and Dan Smith, eds. 1980. *Protest & Survive.* New York: Penguin Books.

Index

Temperatures
  atmospheric, 156. *See also* Heat,
    from explosions
  body, 188, 264. *See also* Fever
Tests, nuclear, 134, 249–251
  Bikini, 25, 32, 91, 249, 270–272
  Nevada, 36–37, 120–121, 124, 250
Tetanus, 224, 225
Tetanus antitoxin, 182, 193
Tetracycline, 193
Texas, 117, 182–184, 208
Texas City, 182–184
Thermal injuries. *See* Burns
Thermal neutrons, 80
Thermal radiation, 24, 119, 126–131,
    141, 155, 157
  See also Burns; Fires; Heat, from
    explosions
Thermonuclear weapons, described,
    31–33
Third-degree burns, 34, 71, 128,
    142–146 passim, 167, 175, 205
Thirst, 44, 188, 285
Thought distortion, 2–3, 308, 309
Thrombocytopenia, 71, 78–90 passim,
    187, 274
Thyroid disorders, 260, 271
  cancer, 103–104, 214, 243, 271
  from radioactive iodine, 158, 161,
    243, 260
Time factor, in decision making, 12
Tinnitus, 78
Titchener, James, 318
Tobias, C. A., 265
Todd, P., 265
Tokunga, M., 104
Tokyo, 131, 173–177, 199, 242
Tokyo University Team, 98
Tomishige, Yasuo, 96, 185
Toxemia, generalized, 269
Transportation, 148, 193–196, 218, 234,
    241, 304
Trapeznikov, Nikolai N., 318
Traumatic injuries. *See* Blast injuries
Treatment, 9–10, 148, 179–245 passim,
    272–278, 305–306
  See also Hospital resources; Medical
    personnel
Triage, 277–278
  for burn patients, 206, 208
  defined, 186, 205, 260
  in Detroit scenario, 148
  psychological capacity for, 200
  and radiation injuries, 273
Tritium, 31, 32
Tsipis, Kosta, 28–39
Tuberculosis, 156, 225, 226–229, 239
Tulinov, Vladimir B., 318

Tumors
  benign, 156
  malignant. *See* Cancers
Typhus, 220, 221, 225

Ultraviolet radiation (UV-B), 156,
    282–285, 303
Uncertainties, in calculations, 114–118
Unconsciousness, 78
United Kingdom, 186, 275, 305
  nuclear-attack scenario for, 163–172
  and tuberculosis, 228, 229
  in World War II, 130
United Nations, 247
United States, 12, 111–136, 303
  blood supply in, 193
  firestorm possibilities in, 131
  hospital resources in, 9, 128, 146,
    148–149, 182–184, 191–193,
    206–210, 223, 238
  medical research of, 93–94
  nuclear arsenals of, 8, 33
  nuclear-attack scenarios in, 115, 118,
    133–134, 137–150, 186, 189–196,
    211–232, 243, 287–288, 296
  oil-refining capacity of, 117, 218
  and ozone content, 282
  and Philippines, 198
  war dead of, 11, 17, 152
  See also Tests, nuclear
Unreality, feelings of, 50–54, 56
Uranium, 24, 27
Uranium-235, 27, 29–30, 187–188
Uranium-238, 27, 31, 32
Urination, increased, 285
USSR. *See* Soviet Union
Utah, 251

Vaccination, 214, 225
Vartanyan, Marat E., 318
Vegetation, 57–58, 303
  See also Crops
Vehicles, 148, 196
  See also Transportation
Ventilation, in shelters, 220
Verheggen, William, 318
Vertigo, 78
Vienna, 227
Vietnam War, 193
Vitamin D toxicity, 284, 285
Vitamin shortage, 19
Vomiting
  from grave digging, 198
  from radiation exposure, 58, 80, 81,
    86, 170, 186–187, 236, 263, 264,
    271, 273, 277
  from vitamin D toxicity, 285
Voors, A. W., 215